SpringerBriefs in Architectural Design and Technology

Indexed by SCOPUS

Understanding the complex relationship between design and technology is increasingly critical to the field of Architecture. The *Springer Briefs in Architectural Design and Technology* series provides accessible and comprehensive guides for all aspects of current architectural design relating to advances in technology including material science, material technology, structure and form, environmental strategies, building performance and energy, computer simulation and modeling, digital fabrication, and advanced building processes. The series features leading international experts from academia and practice who provide in-depth knowledge on all aspects of integrating architectural design with technical and environmental building solutions towards the challenges of a better world. Provocative and inspirational, each volume in the Series aims to stimulate theoretical and creative advances and question the outcome of technical innovations as well as the far-reaching social, cultural, and environmental challenges that present themselves to architectural design today. Each brief asks why things are as they are, traces the latest trends and provides penetrating, insightful and in-depth views of current topics of architectural design. *Springer Briefs in Architectural Design and Technology* provides must-have, cutting-edge content that becomes an essential reference for academics, practitioners, and students of Architecture worldwide.

More information about this series at http://www.springer.com/series/13482

Mahmud Rezaei

Reviewing Design Process Theories

Discourses in Architecture, Urban Design and Planning Theories

Springer

Mahmud Rezaei
Architecture and Urban Planning Faculty, IAU
Central Tehran Branch
Tehran, Iran

Visiting Scholar
City Institute
York University
Toronto, ON, Canada

ISSN 2199-580X ISSN 2199-5818 (electronic)
SpringerBriefs in Architectural Design and Technology
ISBN 978-3-030-61915-2 ISBN 978-3-030-61916-9 (eBook)
https://doi.org/10.1007/978-3-030-61916-9

This Springer imprint is published by the registered company Springer Nature Switzerland AG
The registered company address is: Gewerbestrasse 11, 6330 Cham, Switzerland

Foreword by Manuel Gausa

Theory as (New) Criteria of Action

In the case of the present publication, *Reviewing Design Process Theories* of Prof. Mahmud Rezaei we are in front of a theoretical linked construction capable of building a story line that defends the importance of a new thinking construction and "grammatical" approach in front of the exaltation of the notion of sacred historical episodes of our architectural and urban discipline that we have witnessed in the past decades. It unfolds the connection in between a new structure of though and conception with a certain re-combination of the continuous and dis-continuous concepts in architecture, particularly in the case of the problem of the language and typological periods. Moreover, the theoretical approach responds to a new operative logic which underlies the new culture of our time and that could be recognized in several historic and contemporaneous experiences that are transversally connected with a multi-layered reading of the reality.

The will of research of this interesting work is related to several concerns that today are decisive in order to face—from new theoretical referents through past knowledge totems—the big methodological and disciplinary transformations that we have been experimenting in the last 20 years.

It is specially illuminating the defence of a close connection in between on the one hand the assimilation of parameters of connexion and taxonomic crossing and overlapping interests associated to the past and with contemporaneous project, and on the other hand its relation to key elements of transition that are implicit in the configurations that empower certain spatial processes which are present either in the architecture here displayed and in its cultural translation.

The great ambition of the work of Mahmud Rezaei is the one of having constructed not just an operative taxonomy, but an authentic atlas of sequences that refer not just to the imaginary based on the own universe developed, but to the new referential apparatus that this logics could built and constitute by itself in relation to a new cultural action. Beside its value as a theoretical work, the work shows the opportunities that this new thinking approach could have in terms of architectural design, not just in terms of the architectural element, but also as a typology of production of the architecture and urban space themself.

The division in the different descriptive parts that structure this work are particularly convincing, and permits, in a scientifically and bibliographically well argued way, to go over an implicit cartography unfolded, based on a common recognizable references that are approached with new interpretative keys. In fact, the compared methodology, the analogic study and the interconnection in between processes and significant dynamics (sequences, series and recursive phenomena) has build the base of a work directed to combine selective analysis and underlying interpretations that are presented as latent and/or emergent tendencies.

Reflections thought as a virtual "instruction manual" for "involved" or "implicated" architects that believe in the importance of a creative culture: instruction as "knowledge", as "learning", but also as "education" and as "vector of progress orientation" ... in every sense of the term.

Reflections that not only raise interesting concepts around the spatial formulation and its necessary qualitative definition, but also a certain propositional, optimistic and operative ethics at the same time, in relation with the construction of our own qualitative habitats (ultimate purpose of architecture).

There would be a lot of disciplinary responsibility, but also of noble, healthy and involved ethic in this document of Mahmud Rezaei.

The last decades have, in effect, led to the evidence of a spectacular leap of scale in the definition of our spaces of relationship and coexistence—of our own habitats—connected with the own increase in mobility and distance communication, of the relocation of exchanges and the technological and material transformation capacity of our environment.

But also, with the complex and digital processing of information. With all its benefits and conflicts, potentials and implied threats.

Our challenge, as architects, is to propose new formulations of space (and architectural project) in synergy "in" and "with" a real, virtual and vocationally more complex medium.

Or if you prefer, more open to complexity...

In effect, the notion—so important today—of complexity alludes to this: to the ability to combine and conjugate—to activate and to interact—multiple and not always harmonious layers of information, analyzes and processes in a same infrastructural framework of (inter)relationship.

The old craft gives today the floor to the Research (and therefore the transmission of certainties and ritualized trues to the construction of criteria ... criteria of action).

Beyond old academic strategies focused on methodological liturgy and competitive training, the new theoretical and scientific researches must be capable of generating new valences ... and new values; cultural and social.

Conceptual intention and intellectual position: ethical and purposeful (propositional).

Today we need to generate not only knowledge but a creative energy and a positive shared stimulation Celebrating architecture and enjoying research ... that should be a possible common discourse today, fundamental to claiming—and projecting—the transversal role of the Architecture as a very old and new (trans)discipline.

This book enthusiastically contributes to these premises, establishing itself as a serious and enjoyable publication at the same time; lucid and playful; full of cultured quotes and intense aphorisms; of deep messages and witty phrases: in an original "hybrid" combination between "self-help book", "instruction manual" and "theoretical essay".

Barcelona, Spain/Genoa, Italy
July 2020

Manuel Gausa

Foreword by Emad Mushtaha

I share with Dr. Mahmud Rezaei his interest in understanding design theories and therefore, having read his book, *A Review of Design Process Theories*, I was delighted to write this Foreword to it. I believe that this book will provide an effective learning experience and a referenced resource for both professionals and students and I very much hope that it will.

Before they design a project, architectural students and designers find that the most frequent and challenging questions that always arise are how to start the design concept, where to find design ideas, and how to generate and improve them. Design in general is the platform connecting all professions, and the architect is probably the only maestro to bring so many professionals together. Since architecture is complex and impinges on many other professions, architects must deepen their knowledge and understanding of science, engineering, technology, philosophy, theory, history, sociology, social, materials, environmental issues, and the economy, amongst others. In past years when architects designed buildings or other things, they had many philosophical reasons for every single line in their drawings. The works of recent architectural students and professionals who are aware of the past make us realize that current design either evolves from or sometimes plagiarizes the designs of recent buildings or objects. It seems that academia do not provide enough design theories and processes. Therefore, there is an urgent need for designers and educators to pause and rethink how they can reform the current situation for designers and find ways to nourish brains with design theory and philosophy.

This book comprehensively and consistently finds such ways. It categorizes design theories in eight discourses with a massive literature review, tables, and diagrams that mention related thinkers and describe their contributions. Several theories and logical processes connected with design in general—such as the style of design thinking, possible directions in the design process, the roles and position of designers, the tools for designing, the paradigms and scope of design—are extensively covered for the reader's edification and the disciplining of his thought. The writer has made a great contribution to the field, of much importance, above all to students, researchers, academics, and professionals.

The organization of the book does not try to fix or finalize the categories and subcategories of design. Instead, the writer calls on interested scholars in their future studies to complete, change or modify, and develop his themes. This flexible approach and openness allow us to go further to complete his concepts and link them to other methods, techniques, styles, and tactics, rather than to dominating theories on the process of design.

In this new age of global interconnectivity and interdependence, I have found that The "Review of Design Process Theories" has the solution for combining past and recent design theories with the insights of philosophers and scholars from many perspectives in a sort of review of all the design literature. Although the book is short, it gives readers the chance to take a quick look at many types of theory. This leads me to conclude that, because of its unique approach and writing style, everyone interested in understanding design theories should have this book in their library.

Sharjah, UAE
July 2020

Emad Mushtaha

Foreword by Graham McKay

Theory is a conceptual framework or coherent set of ideas and models that aims to describe, explain, predict or recommend a phenomenon.

Books such as this one that bring together design process theories are especially welcome because, for all the things in the world that are designed, we know very little about the process by which they were.

Students, practitioners and online media generally give theory a wide berth because it's impractical, text-dependent, abstract and not about images. This is unfortunate but theory can't help being theoretical. Architects may write extensively about their projects but it doesn't count as theory if they don't shed any light on what happened between the beginning and the end of the design process. This in-between bit is often called a *black box* because it functions to isolate and contain all those unknowable parts of the design process but it remains to be proven if they're unknowable because they defy language and understanding, or because we simply don't want to open it and see what's in there. Either we're comfortable believing the design process is mysterious and awesome, or we're curious about how those inputs become output.

I'm on the curious side of the fence. If we want to believe design decisions are something more than some designer's arbitrary preferences, then they must be capable of being explained as a sequence of discrete steps with the givens at one end and the thing designed at the other. To invoke the artist's defense and say "this is how I wanted it to be" puts the entire process in a black box and means the author is either ignorant of the mechanism or does not want to divulge it. The stance of this book is that the design process can be understood and it gives us eight different ways of understanding it.

Many of these involve elaborating the two steps of analysis and synthesis into some or all of (1) Define the problem, (2) Collect and analyze information, (3) Synthesize it, (4) Evaluate it, (5) Optimize it and (6) Communicate it. Whatever happens, it's still recognizable as the design process as simulated in university architecture departments around the world. It begins with students being asked to gather information on a particular typology and to then present it as "case studies" that stand in for knowledge and experience. The expectation is that students will then synthesize that quasi-knowledge and pseudo-experience and solve the design problem.

The relative importance of "process", "outcome" and "communication" will vary according to country, institution, course and instructor but all share the same assumptions, namely, that the knowledge and experience will be appropriate and sufficient, that all that *can* be gained from it be gained from it, that its relation to the problem is understood and, crucially, that it will be *creatively* synthesized into a solution. Even this crude simulation contains the black box of creativity—that thing no-one knows how to define let alone teach.

On the other hand, design processes for which *all* relationships between the inputs, connections and outputs were known would be able to process any givens into a suitable outcome. These exist. One example is the kitchen store that will design you a kitchen if you give their designer some dimensions, a list of wants and a budget. Here, the givens are given, the components are known, and which ones are selected and how they're put together is a matter of functional requirements, preferences and budget mixed in with some salesmanship. An infinite number of designs is possible, but they will all follow the same process, including feedback and iterations. It's a perfectly adequate process for kitchen design, as it also is for many types of prefabricated buildings. Designing in a contextual neoclassical style isn't that dissimilar since a reproduction building in some historically sensitive context will have the stylistic pieces—the "vocabulary", some like to say—the process, and the end result all largely predetermined. However, many people prefer to believe that creativity in the design process is something more than an infinite number of combinations of parts. The problem here is that anything beyond that is going to work against the economies of modularity and prefabrication. The marriage of parametric design with algorithmic prefabrication claims to square this circle but only by defining architectural creativity in terms of what prefabrication and construction processes can accommodate. Prefabrication and parametric design approaches are thus alike in shifting the goalposts by narrowing the definition of creativity to suit.

Theory and language are inextricably bound but if we're ever going to get a look inside that black box, we're going to need some language to describe what we find. A design process can be problem-oriented or solution-oriented. It can be analysis-oriented or synthesis-oriented. The relationship between problem and solution can be linear, parallel or cyclic. The emphasis can be on procedure—*The How*, or on product—*The What*. Designing can go from inside to outside or vice-versa. It can be top-down or bottom-up, and whole-to-detail or detail-to-whole. It can be participatory or prescriptive. Synthesis may follow analysis or vice-versa or it might proceed in parallel—assuming they're even separate steps to begin with. This is not complicating things for the sake of it. We need this language.

It was no surprise to learn that solution-based propositions are more common in Architecture and problem-based propositions more common in Science. Architecture may have a black box stage known as "Illumination" or "The Creative Spark" but we know from Archimedes that Science has its "Eureka moments" when analysis is suddenly synthesized into an elegant proposition. It's still the same black box. One important difference is that problems in architecture are notoriously more fluid than in science. Paul Rudolph once said the buildings of Ludwig Mies van der Rohe were successful only because Mies chose to solve so few problems. It is true that architects tend to solve the problems they set themselves.

The book collects the results of much research and, if one is looking for confirmation bias, there's something for everybody. I smiled when I read that scientists endeavor to discover the structure of the problem while architects prefer to produce a series of superior solutions. I learned that risk can be lessened by breaking a problem down into many smaller, solvable problems and to then synthesize those solutions. I understood this as analogous to learning how to play a difficult piece of piano music where you master the notes and phrasing for each hand separately before you try playing with both. I nodded when I read that designers who draw upon their experiences and memories often produce better results that those that follow a methodology. I don't take this as proof that intuition is preferable for intuition is simply a different word for black box. We can't assume something is *intuitive* simply because we don't understand how the brain stores and filters data and makes connections between it. Perhaps if we looked more at history and experience as a huge resource of case studies of what worked and what didn't work for a given situation or problem, we might be able to index and store this data so it can be more readily retrieved when needed. The entire history of the built environment provides examples of specific outputs from specific givens so perhaps if we indexed history according to what problems were solved, we might be able to better infer the workings of this black box. The "Eureka moment" might turn out to be nothing more than the endorphin moment when the brain overrides poor indexing.

If we can look at the built environment as the sum of past problems solved, then it's reasonable to ask what problem is this book solving? More to the point, what problem is design process theory attempting to solve? What is *The Why*? I had no answer until close to the end where I learned we are currently going through a phase of participatory, bottom-up design, presumably after a period of prescriptive, egoistic design. That there could be such a switch suggests the two might just be different ways of achieving the same thing. But what? It could represent a shift towards the satisfaction of users and away from the satisfaction of clients but, even if that doesn't turn out to be the case, it still points to some meta-goal shared by all design processes and that gives both process and end result meaning.

Graham McKay
School of Public Architecture
Michael Graves College
Wenzhou-Kean University
Wenzhou, China

Acknowledgements

Since 2000, the experiences, feedbacks and reviews received from students as well as the colleagues and friends in the fields of arts, architecture, urban design and planning encouraged me to improve, firm and coherent my previous writings into this current English version. Therefore, I would like to express my deep gratitude to those who paved the way for me.

Not only various branches of my based university in Iran and the United Arab Emirates (IAU) but also some institutions including the University of Sharjah as well as City Institute at York University in Toronto reinforced me to author and publish this book globally.

In addition to the institutional supports, Prof. Linda Peak, the director of the City Institute, boosted my thoughts with her wise recommendations and connecting me to the right people. Professor Ali Asgary, at Faculty of Liberal Arts and Professional Studies, York University, also read the primary versions of the book and hinted me with his deep knowledge and guides. Douglas Young, associate professor and coordinator of urban studies program, department of social science at York University, suggested me fairly how to modify some parts.

Discussions with Assistant Professor Arnisson Andre Ortega at the University of the Philippines, has surely affected my thoughts. Leeann Bennett and Francesca D'Amico-Cuthbert were among those executive committees at City Institute who provided me with a welcoming and professional networking environment. I appreciate them for letting me have proper facilities and spaces to do my studies.

Feedbacks, fact-checking and constructed comments from the following professors helped me to advance the book from the very basic notions into the final stages for the production:

- Graham McKay, Assistant Professor at School of Public Architecture, Michael Graves College, Wenzhou-Kean
- Prof. Manuel Gausa, Planning and Urban Design—DAD UNIGE, Genova, Barcelona, Spain
- Dr. Emad Mushtaha, Associate Professor, Departments of Architectural Engineering, University of Sharjah, UAE
- Prof. George Katodrytis, Head of Department, American University of Sharjah.

I would like to thank my linguist friends Dr. Hamed Abdi and Hamed Hosseini for finding appropriate equivalents for some specific words in diverse languages. During the book production, I learned from Juliana Pitanguy and Amudha Vijayarangan, that patience may have no limits as they helped me out more than what I had expected.

Drawing and sketching have all been completed with the help of Mohadese Hamidi, Moslem Mostajeran, Shaghayegh Shaybani. I am, therefore, grateful to all who have cooperated in this book. It would have been difficult to complete this research without them. None of the mentioned people, however, are to be blamed for the shortcomings of my work.

Mahmud Rezaei

Introduction: Design Process Theories and Concepts

Since 2003, the author of this book has worked on design processes and tried to find a meta-theory in which he would be able to view most type of theories. Various classifications are prepared that are presented in some texts (Rezaei 2003a, 2004, 2011, 2014a). But this book is the newest version with more reviewed texts.

This effort may complete the views of the world of design theories, however. The book has tried to see the design theories from various perspectives in order to make it easier to locate new looks. It tries to go beyond the boundaries of design disciplines and activities to flourish noble ideas. That is why it aims at generating a meta-theory and blurs the borders with systematic and deep search through history of design.

Meantime, it seems as brief as possible in a sort of design literature review. Even though almost all the mentioned texts in the book have been reviewed during a long period of time, it is still a short book to give the chance to the readers to have even a quick look at theory types.

Design method and the creation of space and form, especially in architecture, urban design, planning and related built environment fields, have always been among the most important challenges. Many believe that there is no specific method of designing in this profession at all. Moreover, in defining any process, following an assumed hierarchy is not guaranteed throughout the stages of that process. A large group of designers and critics have even claimed that principally, there is no feasible definite method of space design and that the methods in the books are only techniques to harness the human thoughts in specific stages of that path, and following it would not necessarily be effective.

However, other paradigms make opposite claims and even try to define the empirical, intuitional and artistic aspects of design in the form of a defined process. If a space is designed, a method must have been used, a method which has found a solution for the problem of designing. Some experts believe that the mentioned method can be researched and turned into a process. While some scholars are relating the 'scientific' problem-solving process to the spatial design such as architectural design process, others emphasize on the fundamental differences between the spatial design process and design processes in other disciplines such as sciences, engineering, computer, management and planning. Based on this view, the spatial, physical, behavioral and conceptual aspects in architecture and urban design have simultaneously turned space design into totally another type of problems, which common problem-solving approaches are not capable of resolving (Rezaei 2003a).

For instance, the tools for creating space works like plans, sections, façades and sketches or computerized 3D models are among those cases that differentiate spatial design from other design processes. To many, these tools are effective in the design process, being able to change the products of form and space. The thing that most scholars of the field of design process are aligned with is the presence of at least two stages of analysis and synthesis in this process. The design process, for many, moves from analysis toward synthesis, in an inductive and bottom-up direction while some deem it otherwise. To the second group, the space design process has a top-down direction. Of course, there are also middle-of-the-ground views that point to simultaneous convergence and divergence in the design process.

Besides these three views on the 'process' of the design, there are theories that recognize paradigms and patterns for architectural design. By analyzing architectural works or through interviewing designers, some have managed to categorize methods used in the history of architecture. For opponents, the threat of such paradigms is an emphasis on product orientation rather than process orientation. Furthermore, it may diminish the stage of cognition in the design process. Moreover, artistic creativity can be undermined when following a pre-defined paradigm.

The postmodern and poststructuralist critiques of modernism have given new directions to the design method. Postmodernism has underlined historical or typological associations in the design process. Post-structuralism has highlighted the meaning in factors that are lodged in the physical topography of the site or the architectural configuration of the environment. The important point is not thus to start with a functional program. The cultural, social and political roles and implications of each role, in other words the 'context' in general, must be carefully considered (Hearn 2003: 323–325).

By reviewing several theories and their categorizations in different fields, the author has addressed theories and concepts in designing process within the domains of built environment, architecture, landscape, urban design and planning. Harison Fraker, Alex Krieger, John Punter, Paul Davidoff and John Lang are among scholars who have classified the design theories. But in this book, the author has tried to set his classification based on the two main phases of design procedure, namely analysis and synthesis. These two steps have normally been accepted to play a basic role in design methodology. As such, the two have been considered as the main criterion to classify the design theories in this book. Moreover, the way, direction and procedure, a designer approach to these two, may vary from focusing on product/output to emphasizing on process/outcome. However, designers may also take different thinking positions. One might take a rational approach by working at the office and using typological method, whereas some may visit the site and use their various experiences to improve the relevant analysis and synthesis steps. Furthermore, the author has categorized many original texts related to the mainstream of design. By in-depth studies, the selected texts have been classified and sorted out chronologically. The tables are open to accept more ideas for future studies.

Using theories and the categories may facilitate the way of thinking, overviewing, understanding and observing the world. Theory is a conceptual framework, or coherent set of ideas and models, which aims to describe, explain, predict or recommend a phenomenon. The relation between research and theory is shown in the following diagram. It demonstrates how the science includes knowledge, knowledge consists of theories, and the main components of the theories are prepositions that contain concepts.

A researcher tries to fill the significant gaps in the body of knowledge based on theories. Therefore, theories organize the thoughts and formulate the research. It supports to understand the meaning and lead us to improve human knowledge and sciences. Having a category of theories will optimize the use of time during our thinking process. It also provides an appropriate base for reading, observing, writing, balancing and creating ideas and concepts which are necessary for the process of any design.

This book categorizes several theories and thoughts relevant to general design process such as the style of design thinking, the directions in design processes, the roles and positions of designers, the tools for designing, the paradigms and scopes of design. Therefore, the book does not focus on a specific theory of design or particular discipline. Rather it classifies the theories mostly dealing with the spectrums of ideas. Each theory has explained briefly in a certain discourse with several examples including texts, theoreticians and main range of concepts which are sorted out chronologically.

Even though the reader will find eight types of theories in this book, an interested scholar may complete, change or modify the categories and subcategories for future studies. There are several topics and questions that the audience, including the author, may address that the book has not covered at all. Further research may deal with the methods, techniques, styles and tactics rather than theories in the process of design. It is, however, necessary to work on the relation between theories and methods (Diagram 1).

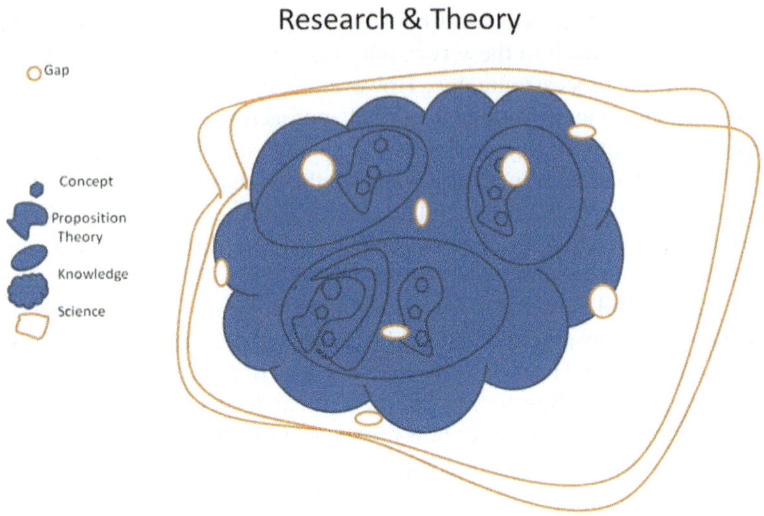

Diagram 1 Relation between research, theory, knowledge and concepts. *Source* Author

Contents

1 Rational and Empirical Theories 1
2 Procedural and Substantive Theories 11
3 Normative and Positive Theories 25
4 Design Scopes 31
5 Design Participation Theories 39
6 Paradigms in Form and Space Creation 47
7 Efficient Tools and Sources in Design Process 59
8 Place and Non-place Theories 69

Conclusion 75
Bibliography 77

Chapter 1
Rational and Empirical Theories

Systematic and Cliché or Intuitive and Creative Approaches

Until the modern period, design theorists seldom feel responsible for coming up with a methodical design process. The strategy to reach a methodical or rational design by Jean-Nicolas-Louis Durand was established in his book in 1802 titled *Précis des leçons d'architecture données à l'École royale polytechnique*. The next record on this topic was by Viollet-le-Duc when he talks about a step-by-step design creation in his 1875 book titled *Histoire de l'habitation humaine, depuis les temps préhistoriques jusqu'à nos jours*. To him, designing was not designer's playing with forms, but rather it responded to the patron's needs, client analysis of the functional 'program' and the 'site' choice. As of the late nineteenth and early twentieth century, inspired by the Art Nouveau movement and modernism, the synthesis of engineering and art gains more strength and the effort to emancipate designers from the past regulations becomes a common practice (Hearn 2003, pp. 179–189). Most urban planning schools either originate from the holistic school or the same comprehensive rational planning or they were the outcome of opposition or reformation of that school. This school systematically looks at issues and connects targets with sources with an emphasis on employing mathematical and quantitative methods (Hudson 1979).

Research on architectural design methods from the aspect of form and space creation is a practice that dates back to the post-1945. This does not mean that there were no enquiries about design methods earlier or that there was no explanation of the methods. Rather, it means that serious thinking on this topic and an all-inclusive, comprehensive, smart, conscious and systematic study on the topic, intending to reveal intuitive, mysterious and, the so-called, black box-like aspects, is more related to the twentieth century and afterward. Until the first half of the twentieth century, most beliefs were based on the vagueness and empirical nature of the design process. For instance, Rainer Maria Rilke, the Bohemian-Austrian German-language poet and novelist, wrote a note on this issue in the early twentieth century, inspired by Paul Cezanne's paintings:

> Ideally speaking, a painter, or generally an artist, should not be aware of his or her insight [while designing], distancing from the aberrations rising from conscious ways of

M. Rezaei, *Reviewing Design Process Theories*,
SpringerBriefs in Architectural Design and Technology,
https://doi.org/10.1007/978-3-030-61916-9_1

> thinking. The proceeding steps of artists, which are mysterious to the artists themselves, must quickly turn into works [of art]. The artist cannot recognize these steps at this moment of transition(Rainer 2002: 66).

Therefore, according to Rilke, the process of producing a work of art will remain unknown even to the very artist. On the contrary, in the same years, there were individuals who sought to make 'design secrets' methodical; for example, Buckminster Fuller began his comprehensive studies on the science of designing in 1927, and he started to put them into practice in 1950, when he started teaching a wide variety of audiences (engineers, industrial designers, materials specialists and chemists) at MIT (Dubberly 2004: 40).

John Chris Jones divides design methods into different fields across several eras. First, there is the 'Craft Evolution Era', in which the craftsmen had rules in their minds. This practice was common since pre-Renaissance. With the help of their limited tools, they created their works by means directly connecting their minds and their hands. The design method in this era was limited to the tools and the mind of the designer, and it was passed down from one generation to another. During this era, small solutions are entangled in the design options. These options were evaluated, and ultimately, the final selection was made.

Jones believes that the next era is the era of 'Designing by Drawing', which lasted from the Renaissance to 1950. In this era, the clients demanded their desired outcome for the designer and the designer, using various problem-solving methods, found a solution in a black box approach. That is, the designer's method was not clear to anyone, and sometimes, even the designer did not know for sure how the problem had been solved. During this era, besides the stages of the previous era, the issue was more demarcated and broken down. Small solutions of each section were described, possible contrasts among them were cleared, and the final choice was made.

'System designing' and a systematic look at designing started in the twentieth century and somehow related to the military requirements of World War II. After the war, the issue of design method was put to research, and generally, it was between the 1960s and 1970s that a movement emerged in researching the design method. At this stage, designing by drawing was a tool to solve a minor problem and also a part of the bigger system that had been impossible for an individual. Jones cites a few new methods during this new era: brainstorming, synectics, removing mental blocks and analysis of interconnected decision areas (AIDA). Therefore, during this period, besides the stages of the previous era, understanding or changing the structure of the problem had also happened.

As Jones believes, the current era is the era of rapid 'technological changes' or socio-technical innovations. In this era, the number of factors that influence design is much more when compared to the past. A modern method of evaluating the problems and discovering designs at the contemporary era is the principle of strategy switching and letting spontaneous thinking infiltrate into planned thinking and vice versa. This is one of the methods that let multiple arenas of intelligence to simultaneously act together. Therefore, this method is sometimes called 'co-intelligence'. Here, according to Jones, the intuitive and rational methods coexist coincidently

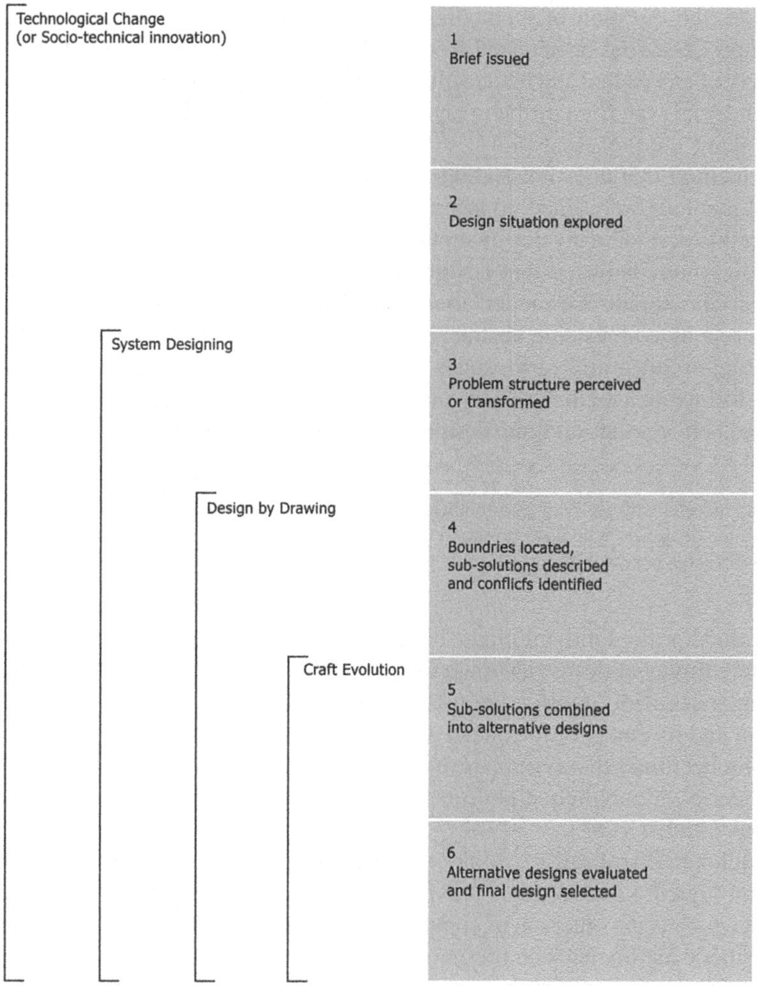

Diagram 1.1 Jones' design process (1970). *Source* (Dubberly 2004, p 57), redrawn by the author

and not exclusively. In this period, the discovery of the situation of the design and issuance of the brief is added to the other stages of the design process. The following chart illustrates it (Jones 1970) (Diagram 1.1).

Based on such methods, important seminars were held on design method in the UK and North America in 1966. Design methods have been influenced by discussions on applying new scientific design methods in different fields, on the one hand, and assiduously holding seminars on architectural design and the aforementioned movement in the 1960s and 1970s, on the other hand, together with the expansion of creativity techniques. With regard to the great importance of the issue of creation in the field of design, the question of methodizing design and making it similar to problem-solving processes has been raised all over again. This is because creativity means the discovery of an original and better solution or the power to find an unconventional and transcendental way for a problem and a new look at reality. If creativity is a process for breaking the presuppositions, so how would it be possible to come up with a design method that does not end up in producing clichés in itself? Creative experimental methods were lined up against rational prototypical methods. On the same basis, at this period, many design methods were suggested to be holistic and general. Therefore, many believed that design process debates in the 1970s were over. For instance, Christopher Alexander, after highlighting the importance of model, pattern and the role of synthesizing abstract patterns with respect to the design background in creating architectural form, states that no one will become a better designer by blindly following a method. He rejects design method as a problem-solving research topic and believes that studying design will separate from the act of design. He added that,

> In fact, people who study design methods without also practicing design are almost always frustrated designers who have no sap in them, who have lost, or never had, the urge to shape things. Such a person will never be able to say anything sensible about 'how' to shape things either.

Alexander's thoughts highlight both rationalism and empiricism. However, he ultimately thinks of design as 'diagrams' or 'patterns' and believes that each design problem starts with an effort to achieve fitness between two entities: the form in question and its context (Alexander 1973, p 15). Similar to Christopher Alexander, Bruce Archer thinks that systematic methods in the computer techniques and management theories greatly help architects assess the design problem and develop design solutions (Archer 1964).

Colquhoun[1] first discusses Maldonado[2]'s opinions that pure intuition should be based on logical knowledge of previous solutions and that the act of creation is a process which is the outcome of comparing the forms established from past needs or past aesthetic ideologies with present requirements. Colquhoun then generalizes this issue to other non-architectural design processes and believes that design process is made up of the challenge between two non-conforming parts: the first being the development process of form based on biotechnological determinism and the second being the freedom of speech and intuitive selections (Colquhoun 1996).

In 1970, Eastman[3] made at least two discoveries about analyzing the process of intuitive designing. The first indicated that there is a clear relationship between design constraints and limitations and the manner of design presentation such as words, numbers, diagrams, plans, sections and perspectives. The second finding was about recognizing the project and its problems, and the designers who put emphasis on the direct retrieval of past experiences and their memories are superior to those

[1] Alan Colquhoun.

[2] Tomas Maldonado.

[3] Charles Eastman.

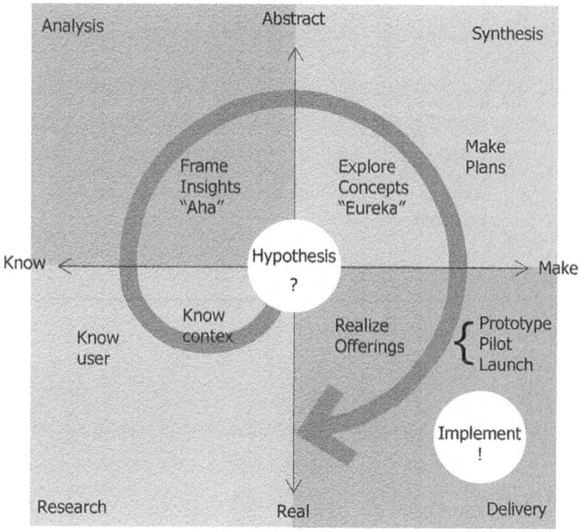

Diagram 1.2 Innovation planning by Vijay Kumar. *Source* (Dubberly 2004, p 125), redrawn by the author

who benefit from external clues existing in the topic for producing the design based on the constraints of the problem (Eastman 1970, p 20). In other words, he puts the concept of intuitive design against methodical design and is of the opinion that having a correct and clear method is not necessarily superior to intuitive methods. In most cases, designers do not think about their act of designing and behave only in an intuitive way. Based on Eastman's studies and similar research results, it can be said that a problem-solving process will not always reach a clear conclusion in 'architectural' design, although the product of architectural design has a clear relationship with its process. In other words, the relationship of reaching a spatial design depends on the designer and is a one-way street.

Another model of design process was introduced by Vijay Kumar at the 2003 HITS Conference (Humans, Interaction, Technology and Strategy) in Chicago. His design process model was called 'Innovation Planning'. In the design process, he spoke of innovation as the jump from '*insight*' to '*concept*'—from *'Aha!'* to *'eureka!'*—describing it as a revelation, magic, genius, intuition and a hunch in designing (Kumar 2003). The magical moment will stay intuitive, and its entity will remain a secret resting concealed inside the black box. Therefore, the realization of the response to the design of space or arriving at the final design may be done rationally or intuitively (Diagram 1.2).

With apologies to those design methodologists who like maps, Bryan Lawson offers the following map. The map depicts the walking process through a diagram. All human know how to do it, but none has learned it this way. This chart, on the one hand, shows the relationship between *rationalism* and *empiricism* and methodization of experience, and, on the other hand, it shows the distance between 'the act of design' with the 'problem-solving process'; the same problem of dancing and sauntering pointed out in the book's Introduction; the same thing that occurs on the basis of

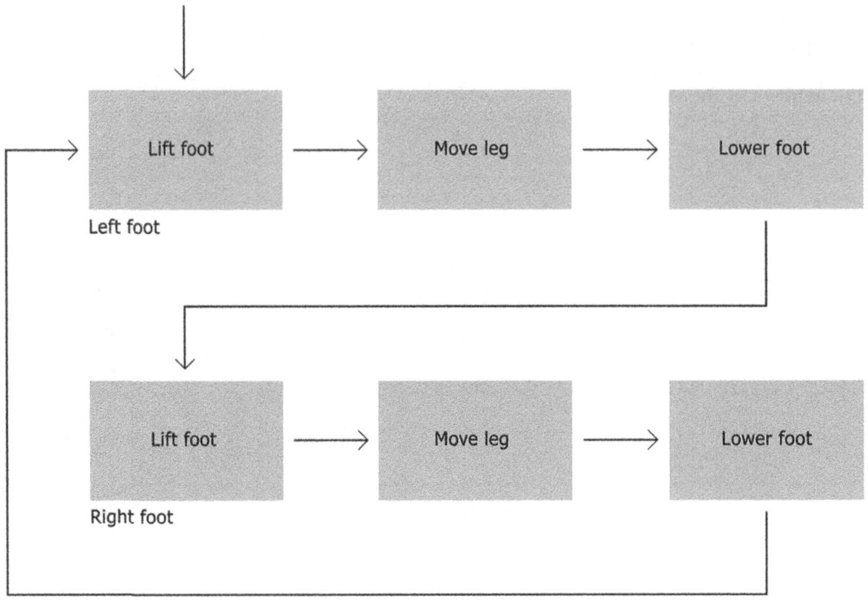

Diagram 1.3 Walking process by Bryan Lawson. *Source* (Dubberly 2004, p 28), redrawn by the author

experience, but can be expressed through a problem-solving process, though not necessarily carried out in such manner (Diagram 1.3).

Rationalism and empiricism can also be studied in urban design. Quoting Geoffrey Broadbent, it has been pointed out three types of major concepts through history of designing: rational, empirical and pragmatic concepts. Broadbent also follows William James in believing that the pragmatist philosophy is a branch of empiricist philosophy. When rationalist ideas move on from the logical stages and to follow scientific analysis and deduction from discussions and primary principles, empirical theories take form through observation and perception helped by an individual's senses and experience. The essence of rational ideas is made of scientific, mathematical and geometrical concepts, and regardless of experience and qualitative essence of objects, it binds them all together. This is while the essence of empirical ideas is based on human senses and observations.

The rational manner of thinking has been interpreted as a set of rational theories used in urban designing. Since the eighteenth century, French architects like Laugier[4] have been trying to look for and discover essential or fundamental architectural elements through pure ration, i.e., the beginning of reasoning from 'primary

[4]Marc Antoine Laugier.

axioms' and a 'logical approach'. Laugier's efforts led to the writing of 'an introduction to typology of forms in architecture' and were later manifested in another architect's works called Boullée. However, no architect can be found more influential on rationalist urban architecture theories than Quatremèr.[5] His suggestions and definitions of the concept of type make up the foundation of recent rationalist theories of Aldo Rossi, and the Krier brothers in urban design. Since according to empiricist theories, the city and urban design are phenomena that are practically used, experienced and evaluated by people, they will have a close relationship with behaviorist sciences. In these kinds of theories, benefitting from various behaviorist sciences such as psychology, sociology, anthropology and even cultural-human geography, the theoreticians have tried to show a proper approach to urban reality. The main concern of such theories is the discovery of the mechanisms of environment and the quality of users' experience of space. In fact, theories under discussion are user-oriented theories that, besides focusing on human behavior in the environment, both at the individual psychological level and at the collective, sociological level also take into consideration the capacity of the physical environment in welcoming and accommodating various activities.

After reviewing design literature, Ashraf Salama mentions three models of design process. The first is the 'intuitive method' which has a 'black box' approach. According to this model, the act of designing happens in designer's brain and away from the logical or reasonable control. In his opinion, pragmatic, iconic, analogical and canonic designs are all various approaches of designing that represent this model. The second one is the 'rational or problem-solving model' with a 'glass box' approach. In this model, even if designers cannot come up with convincing reasons for all their choices, according to this model, they can still be hopeful to find an acceptable solution for the design problem. The common characteristic in this method, cited by Jones in 1970, is defining, in advance, the objectives, variables and criteria before finding solutions. Systematic design and the pattern language are two methods of representing this model according to Salama. The third model is the 'participatory model' which is based on 'community design' or 'action research' approach. Engaging people in the decision-making process is the only way to consider their ideas, values and needs (Salama 1995, p 78) (Table 1.1).

[5] Antoine Quatremèr.

Table 1.1 Rational and empirical theories in design process

No	Date	Thinkers	Texts	Ideas
1	1706 1788–1823	Jean-Louis de Cordemoy Antoine ChrysostomeQuatremère de Quincy	Essaisurl'architecture Encyclopédieméthodique	Typology as a rationalist method in architecture
2	1753	Marc Antoine Laugier	Essaisurl'architecture	An introduction to typology of forms in architecture, which is considered a rationalist method for reviewing form
3	1802–1819	Jean-Nicolas-Louis Durand	Précis des leçonsd'architecture	The design is first manifested in the plan. Then it is transferred to the façade and then section. None of these should a nonfunctional topic outside the sociological hierarchy and typology of the building interfere
4	1873	Eugène-Emmanuel Viollet-le-Duc	Histoire d'unemaison	The rationalist concept of planning and father of modern architectural theories; stages of step-by-step creation: designing is not a formic game of the architect, but a response to the needs of the client. (1) Planning and site, beginning of the making of plan; (2) When the roof is made; (3) After the plan, time for the structure, façade, materials and openings

(continued)

Table 1.1 (continued)

No	Date	Thinkers	Texts	Ideas
5	1964 1973	Christopher Alexander	Notes on the synthesis of the form a pattern language	Rationalist and procedural approaches to design, Key role of diagrams or patterns in designing to show the relationship between form and context
6	1964	Bruce Archer	Systematic methods for designers	Systematic methods in computer techniques and management theories give designers a great help in evaluating the design problems and developing design solutions
7	1967	Alan Colquhoun	Typology and the design methods	Intuitive choices for designers eventually arrive even in mathematical, scientific and rationalist design methods
8	1970	Charles Eastman	On the analysis of intuitive design processes	Systematic design processes are not necessarily better than other methods
9	1970	John Chris Jones	Design method seeds of human futures	Systematic method consists of three stages of analysis, synthesis and evaluation. Intuition and rationality can exist together, without one deleting the other. Three stages of design: divergence, conversion and convergence
10	1979	Rob Krier	Urban space	Typification of urban spaces (squares, streets and their crossings)

(continued)

Table 1.1 (continued)

No	Date	Thinkers	Texts	Ideas
11	1982	Aldo Rossi	The architecture of the city	Geometrical-abstract purity to realize the building and the city in a rationalist method
12	1988	William H. Whyte	City: rediscovering the center	Observing and studying empiricism in urban design process
13	1995	Ashraf Salama	New trends in architectural education: Designing the design studio	Three models for intuitive designing (pragmatic, iconic, analogical, canonic), rational (systematic and pattern language), participatory (action research, community design, humanistic approach)
14	1998	Dan Braha	A mathematical theory of design: Foundations, algorithms and applications	Formal Design Theory is a mathematical theory of design grounded in the branches of information theory, logic, automata, complexity theory, generalized topology and the theory of category which explores the algebraic representation of design artifacts and so.
15	2002	Douglas Kelbaugh	Typology: An architecture of limits	Limits are essential to freedom. Site and programmatic constraints make the design process easier.
16	2003	Vijay Kumar	Innovation planning: modes, tools, uses	In design process, innovation acts like intuition, discovery, magic and genius

Source Author

Chapter 2
Procedural and Substantive Theories

Substantive views in designing elaborate on the physical essence of built environment and products. These views in architecture and designing form and space also include dealing with building materials, structure, geometry, environment and end users. This view is more concerned with explaining the *whys* of designing.

On the contrary, there is the procedural view, which elaborates and explains the process of the built environment creation such as architecture, landscape architecture, urban design and interior design. These theories focus on design methods, how the spaces are formed and how artificial spaces are created. Therefore, these design theories revolve around the '*how*' questions.

The concerns and attentions of design could be further directed toward the design process, the act of designer and his or her choices. Or the other way round, it could point to the final products, sketches or design products. Procedural designers are more likely interested in group designing, while product-oriented or solution-oriented designers gave more credit to ideas.

Substantive approaches conceptualize design as spontaneous actions. This is where intuitive design methods do not care much direct about the 'process' of designing when they arrive at the point of 'Aha' or 'Eureka'. But in procedural approaches, growth and emergence work well among the factors insisting on a 'gradual' or 'process' measure.

Procedural theories aim at describing and explaining the design processes. For that reason, these types of theory should have the ability to distinguish between the products that originate from different design processes. According to Colin Rowe's Collage City, designers create their works by adding to or reducing from the status quo that is a procedure which is more like the art of collage than any other form of arts.

Instead, a substantive theory aims at analyzing design products. It has to do more with the essence of phenomena dealt with by designers in their line of work. The most principal phenomena of their interest include recognizing the essence, quality and manner of the functions of built environment, the capability of the environment

M. Rezaei, *Reviewing Design Process Theories*,
SpringerBriefs in Architectural Design and Technology,
https://doi.org/10.1007/978-3-030-61916-9_2

for accepting and accommodating activities from the physiological aspect and from the qualities of environmental aesthetic experiences (Rezaei 2014a, p. 35).

As a result, the views that consider designing as a design process tend toward gradual measures and deal with the 'how' problem, while substantive views of the design emphasize on spontaneous or immediate measures and delve into the 'what' problems. Attention to generative procedures and evolution has brought about the concept of procedural views in designing. The building is recounting the concept of process, more than just being an architectural product.

Salama has discussed the design process and summarized the different related views in Table 2.1. It is clear that the stages of identifying the problem, information gathering, formulating the program, testing solutions and selecting from alternatives are the essential parts as shown in the table. The student design process, described by Thornley in 1963, does not include any of these activities (Salama 1995, p 86).

Apart from these points, the direction or orientation of the design process is of particular importance. Product-oriented theorists are focused on the answer more than the question, and their design is more convergent. This is while procedural theorists insist upon the stage of analysis and shun spontaneous creation or composition. On this basis, one can categorize procedural and substantive views. Design activity may then vary from a top-down transformation to a bottom-up approach.

What relationship lies between the design product (i.e., form) and its process (i.e., context), the relationship through which one can relate the solution, i.e., the created form and spaces, to the problem? How can a problem-solving approach make the process of reaching proper solution? Is the relationship between the problem and solution direct, parallel or cyclical? What is the way, direction or orientation in a design procedure? (Diagram 2.1).

The transformation during design phases may move from a whole to a detail or from a detail to a whole. The designers might adopt a problem-oriented or solution-oriented approach in their designing practice. In the design process, the designer might focus on synthesis or analysis. Designing might take place from inside to outside or vice versa. Different professions have investigated whether the design direction is bottom-up or top-down. These directions in the design processes are explored in three common parts:

A. Analysis-Oriented and Synthesis-Oriented Approaches

What is shared among different steps of design models, not only in architecture but also in other various fields, is the stage of analysis and synthesis. Bill Newkirk has reminded designers about the gradual change from *emphasis on analysis* to *attention to synthesis*, during the design process according to the Diagram 2.2.

Koberg[1] and Bagnall,[2] who taught in the College of Environmental Design at Cal Poly in San Luis Obispo, point out that to solve a problem, we first break the situation or whole problem into parts for examination (analysis). Second, we reassemble the situation based on our perception of improvements discovered in our

---

[1] Don Koberg.

[2] Jim Bagnall.

Table 2.1 The different views of the design process

Thornley 1963 Student design process	Rittel 1970	Corkill and Guenter 1970	Whitaker 1971	Jones 1970	Moore 1974	McGinty 1979
	Identify the problem	Basic definition preliminary program	Recognition definition	Idea	Problem Identification	Initiation
Accumulation of data	Collect information Analyze information	Investigation and analysis Program abstraction	Preparation Analysis	Information Analysis	Analysis of User Needs programming	Preparation
Development of forms	Develop solutions	Synthesis and developing solution	Synthesis	Synthesis	Design synthesis	Proposal making
	Test solutions	Re-evaluation and modification	Evaluation	Evaluation	Selecting from alternatives	Evaluation
Presentation of solution	Communication and implement		Execution	Optimization	Implementation	Action

Source (Salama 1995, p 86)

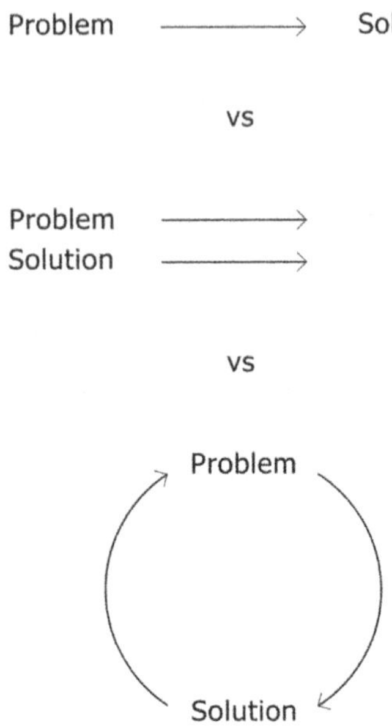

Diagram 2.1 Relationship between the problem and its answer: sequential, parallel or cyclic procedure. *Source* (Dubberly 2004, p 27), redrawn by the author

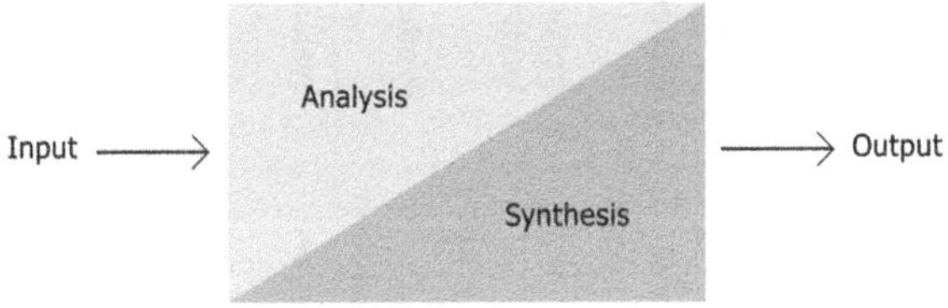

Diagram 2.2 Gradual change from emphasis on analysis to attention to synthesis (by Bill Newkirk). *Source* (Dubberly 2004, p 26), redrawn by the author

study (synthesis). They believe that through analysis, the main theme and guiding idea of the stage of synthesis are extracted. They gradually expanded their archetypal two-step process of analysis and synthesis to three stages of analysis-definition-synthesis. Then they expand it to five stages of analyze-define-ideate-select-implement and ultimately to seven stages of accept-analyze-define-ideate-select-implement-evaluate (Koberg and Bagnall 1990). In the diagram of their working method, the seven stages do not have to follow the next stage. From one step to the next, one can step forward or backward, and there is no restriction to just move forward. These step takings and progress reach the final destination as a result of the interference of factors such as cost, time, energy and the like (Diagrams 2.3 and 2.4).

William Pena and Steven Parshall introduced the 'programming' and 'designing' model specific for architectural design procedure. They suppose the project's

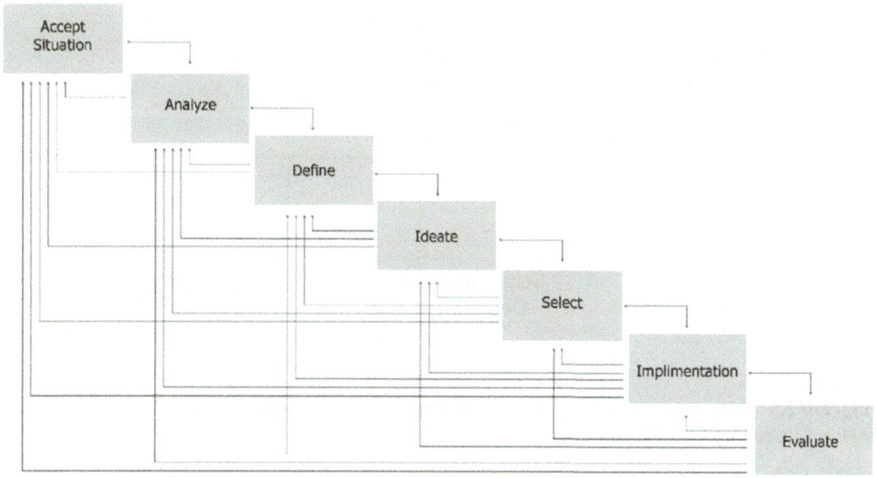

Diagram 2.3 The two-stage procedure of design (analysis and synthesis) is dividable to a three-, five- and ultimately seven-stage procedure. *Source* (Dubberly, 2004, pp 114, 16), redrawn by the author

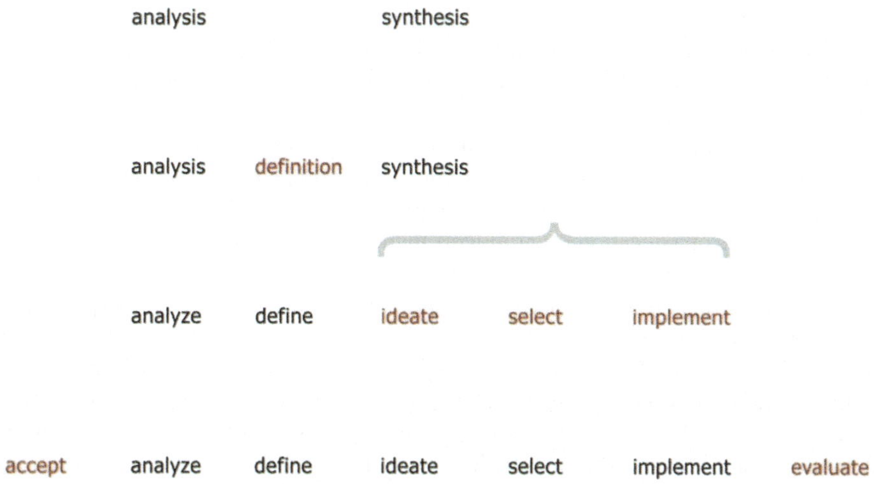

Diagram 2.4 The two-stage procedure of design (analysis and synthesis) is dividable to a three-, five- and ultimately seven-stage procedure. *Source* (Dubberly 2004, pp 114, 16), redrawn by the author

program as problem seeking and the final plan as problem-solving. In other words, they think of programming as equivalent to 'analysis' and designing as equivalent to 'synthesis'. Of course, their attention and the attention of most environmental designers (including architects, urban designers and industrial designers) have been

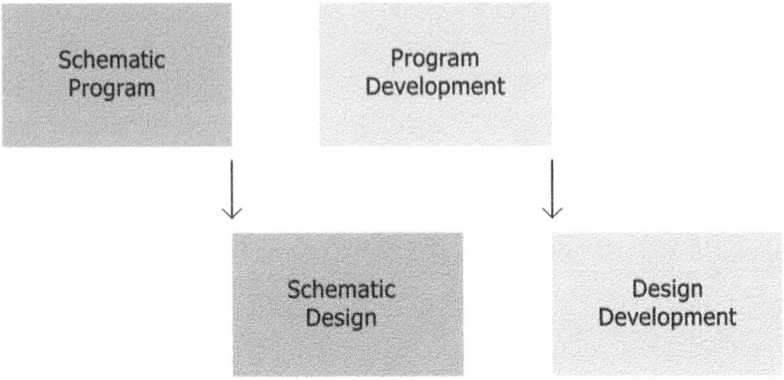

Diagram 2.5 Programming and designing—William Pena and Steven Parshall. *Source* (Dubberly 2004, p 21), redrawn by the author

directed toward synthesis and creativity more than analysis (Parshall and William 2001) (Diagram 2.5).

British psychologist Jane Abercrombie, in her 1965 famous book titled *The Anatomy of Judgment*, used Frostig test to evaluate children's visual understanding and concluded that analytic and synthetic steps, unlike the view of most, are not separated from one another. Bryan Lawson, having a critical view and comparing the creativity process and design process for architects, also stated an important point about the stages of analysis and synthesis. He considers architectural design process as different relevant processes for scientists and engineers and regards the analysis and understanding of the problem in architecture as dependent upon synthesis and production of the final product. In his opinion, the coherence between synthesis and analysis is more of a question for architects than for others, and even so, the synthesis stage might well precede the stage of analysis. While the scientist proceeds with a problem-based approach, the architect gallops on with a solution-based predisposition. He is making the point that finding equivalence between the steps in the creativity process and the steps in the design process. He puts the 'first insight' step from the creativity process on par with the 'problem formulation' step from the design process. He states that the 'preparation' step corresponds with the 'conscious attempts at solution' and likewise the 'incubation' with 'no conscious effort', 'illumination' with 'sudden emergence of idea' and 'verification' with 'conscious development' (Lawson 1980, pp. 148–149) (Diagram 2.6).

B. **Problem-solving and Decision-making approaches**

Akin[3] studies indicate that the design challenge is not only the fact that the problem-solving approach acts weakly in designing, but also the fact that even after determining the goals and recognizing the constraints of the scheme, one cannot suppose them as fixed rules and unchanging paradigms. Therefore, a unique behavior in

[3]Omar Akin.

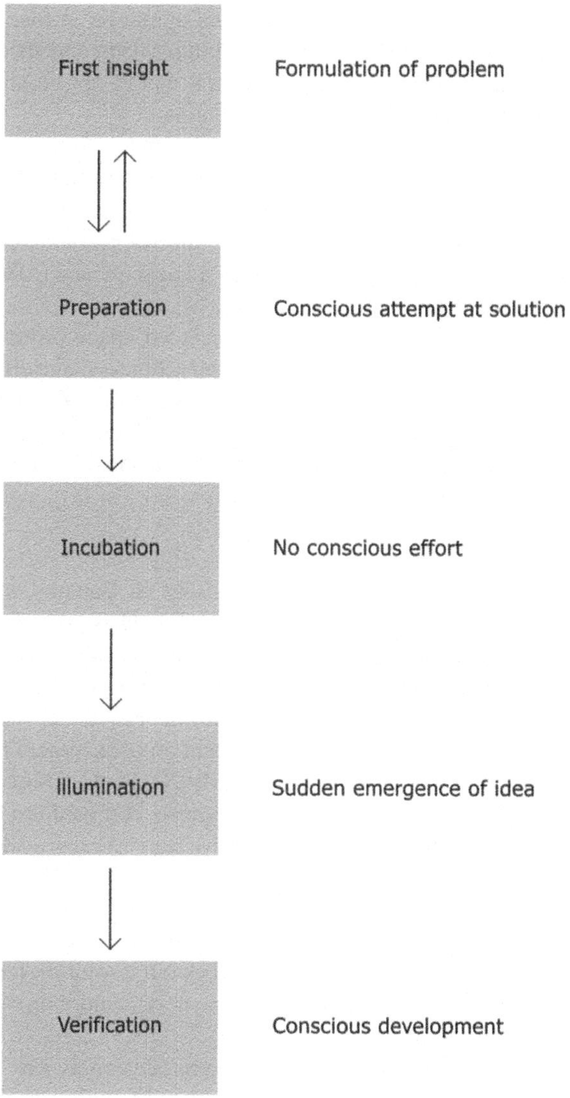

Diagram 2.6 Creativity process and design process—Bryan Lawson. *Sources* (Lawson 1980, p. 149) (Dubberly 2004, p 42), redrawn by the author

designing is regenerating new goals and redefining constraints. By further defining the solutions and progressing the problems of the project, there would be no more reason to preserve the previous goals and demands. The designer is free to change such cases. At a certain moment of working, the designing architect, by means of his or her power of insight, goes beyond the problem: the Aha moment (Akin 1978)!

Analysis is a part of all phases of design, and synthesis is found very early in the process. This means that the problem of designing form and space is a different problem; therefore, the problem-solving approach would not be able to act in the same way as it deals with non-architectural problems.

Schön[4] rejects the specific limitation of the architect to a conceived problem and deems that designers determine the problem of designing themselves. That is, the problem formulation process is a process in which designers name and define their topics of interest and determine framework for the background. This framework enables the designers for their next move and as a consequence makes the resulting discoveries possible (Schön 1983).

One of the differences between design process as space producing and other realms is that in different disciplines of science, the scholars make efforts to discover the structure of the problem, while space designers get to make the proper choice by producing a series of superior solutions. So what the scientists do is problem-based, while the act of creating space emphasizes on the answer and is consequently a solution-based approach (Cross 2001).

C. **Convergence and divergence**

In design process, Christopher Alexander and Bela H. Banathy, similar to Nigel Cross, consider the departure from analysis to synthesis. They point out that analyzing the problem is a divergent and expansive move toward fragmenting or analyzing the issue under scrutiny. Then the reassembly or remounting of those fragments in a new way is a convergent or collective move toward the product or the process result. Bela Banathy highlights the dynamism of divergent and convergent moves and their numerous repetitions during the process (Dubberly 2004). Also Nigel Cross emphasizes on the importance of convergence and synthesis compared to divergence and analysis, and the design process will ultimately have to synthesize analyzed issues and make them converge toward a conclusion and product (Cross 1984) (Diagram 2.7).

Diagram 2.8 depicts opposite of what was said above. It shows that to solve the designing problem, we should first understand and analyze the overall problem well and then break it down into sub-problems and work out a solution for them and then synthesize the solutions. As a result, the emphasis here is more on the problem than on the solution to the problem.

Bela Banathy believes that in the design process, designers first diverge as they account a number of inquiry boundaries, a number of major design options and sets of core values and ideas, and then in making choices and creating an image of

---

[4]Donald Schön.

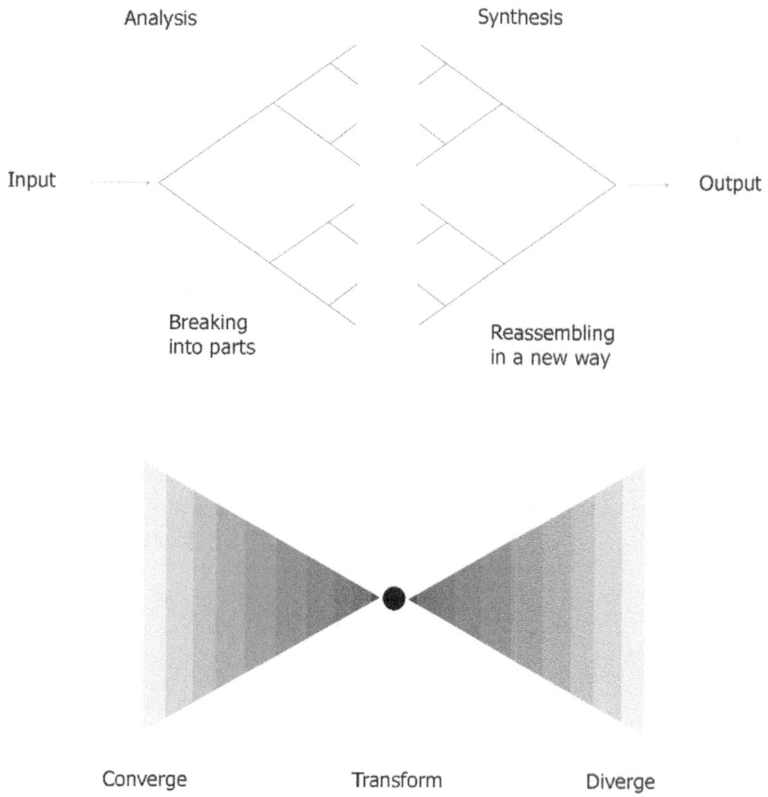

Diagram 2.7 Divergence and convergence versus restriction and expansion. (Dubberly 2004, p 22), redrawn by the author

the future system, they converge. Similarly, the divergent and convergent processes occur when designers are placed in the atmosphere of solution. In every domain of designing (core definition, specification, functions, enabling systems, systematic environment), first by creating various choices for each realm, designers diverge, and then by evaluating alternatives and selecting the most promising and most desirable one, they converge (Dubberly 2004, p 24) (Diagram 2.9).

As Nigel Cross proposes, the ultimate point of the design act is converging on a final evaluated and detailed design proposal. But during the process, some moments are needed to diverge, to widen the search or to seek new ideas and act as starting points. The whole process is therefore convergent with periods of deliberate divergence (Dubberly 2004, p. 25) (Diagram 2.10).

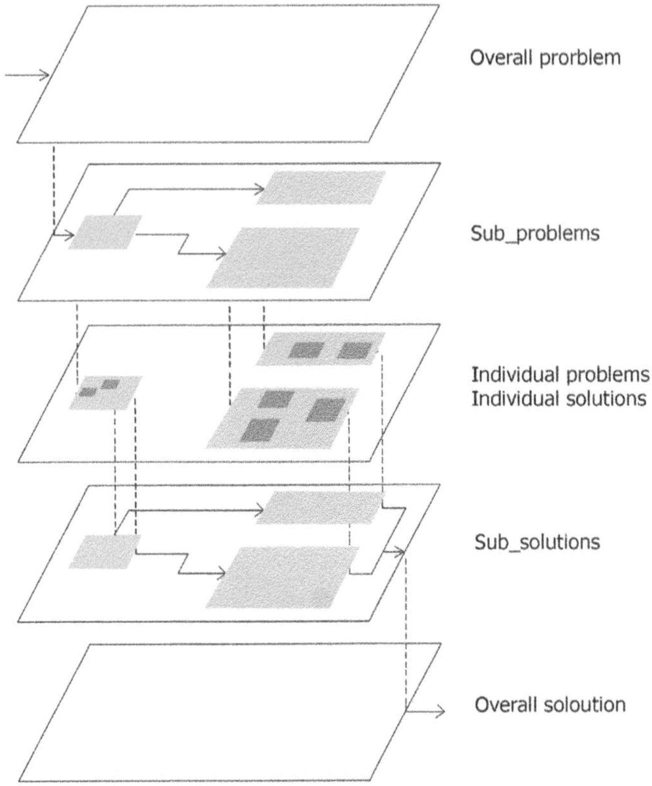

Diagram 2.8 Breaking down and re-composition. (Dubberly 2004, p 23), redrawn by the author

Likewise, the directions in the design methods might be clearly described in concept-knowledge theory (C-K theory). Hatchuel and his colleagues have shown that the C-K theory gives to design theory the similar level of rigor and modeling of decision or programming theory. The two spaces of C (concepts; propositions without a logical status) and K (knowledge; propositions with a logical status) may expand as the result of the mechanisms through four specific operators shown in chart one. Disjunction happens when propositions transform into concepts (going from C space to K space), and conjunction concurs the other way around (Hatchuel 1996), (Hatchuel and Weil 1999, 2002, 2008), (Hatchuel et al. 2013) (Diagram 2.11; Table 2.2).

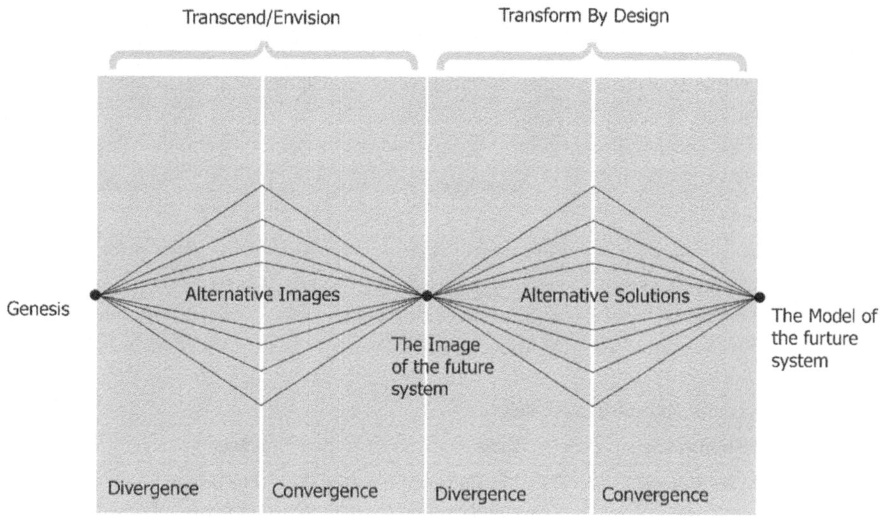

Diagram 2.9 Parallel sequential convergence and divergence in design process—Bela Banathy. (Dubberly 2004, p 24), redrawn by the author

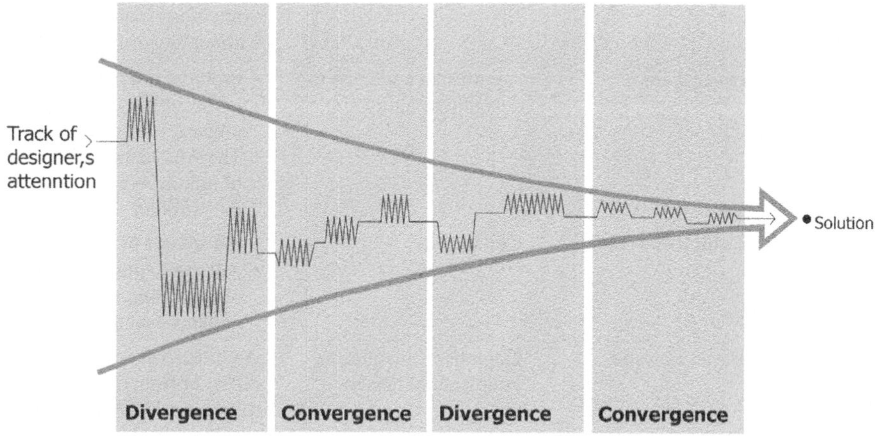

Diagram 2.10 Necessity of design process moving toward convergence. *Source* (Dubberly 2004, p 25), redrawn by the author

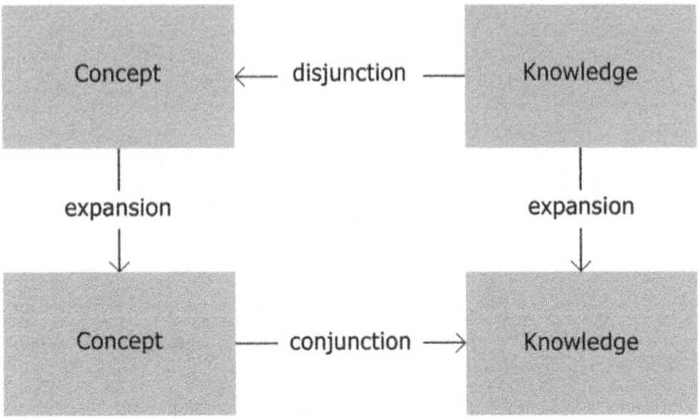

Diagram 2.11 Design square by Hatchuel et al., redrawn by the author

Table 2.2 Procedural and substantive theories

No.	Date	Thinkers	Texts	Ideas
1	1972	Don Koberg	The universal traveler: A vintage guide to creative problem-solving	The creative problem-solving consists of seven stages: acceptance of situation, analysis, definition, ideating, selection, implementation, evaluation (stages work without any order)
2	1978	Omer Akin	A cartesian approach to design rationality	• Problem not being fixed in the architectural design process • The Aha mental insight of architects to go beyond the problem
3	1978	Colin Rowe	Collage city	• An answer to how • Adding to the status quo or reforming it with an emphasis on the past
4	1979	Bryan Lawson	Cognitive strategies in architectural design	Architectural designing is different from other sciences and has different cognitive strategies
5	1979	Jane Dark	Raw materials and architectural design process	The act of design is complicated due to various demands and constraints; however, in most cases, designers tend to get close to a rather simple idea from the very beginning (solution is on the priority, not the problem)

(continued)

Table 2.2 (continued)

No.	Date	Thinkers	Texts	Ideas
6	1983	Paul Oliver	Binary logic in an Islamic city: Isfahan	Study of the methodological process for Sheikh Bahaei' design
7	1983	Donald Schön	The Reflective practitioner: How professionals think in action	Innovations learned during practice help professionals more than analyzing problem formulas learned at school. Designers determine the problem of design themselves
8	1984	Nigel Cross	Design cognition: Results from protocol and other empirical studies of design activity	• The significance of cognition in the design process: In passing from different modes of design solution, the designer should be able to define, redefine and change the design problem • The difference between a scientist and an architect is that the former is problem-based, but the latter is solution-oriented • The superiority of solution-oriented approaches to problem-oriented ones • Lack of only one answer • Convergence and divergence in designing
9	1984	Asghar Talaminaei	Architecture as environment communication (approaches to Semiotics)	• Design method instead of design product • Quality instead of quantity • Notification theories • Urban design processes • Problem seeking or solution finding
10	1985	Hamid Shirvani	The urban design process	• Design methods may be placed into six groups: internalized, synoptic, incremental, fragmental, pluralistic and radical

(continued)

Table 2.2 (continued)

No.	Date	Thinkers	Texts	Ideas
11	1987	Christopher Alexander	A new theory of urban design	• Seven rules of growth (piecemeal growth; the growth of larger wholes; visions; positive outdoor space; building layout; construction rules; and formation of centers) • Tendency to take gradual measures
12	1996	Armand Hatchuel et all	Theories of design (in French)	According to the C-K theory, the two spaces of concepts and knowledge may expand as the result of the disjunction and conjunction mechanisms through four specific operators (between C and K)
13	1998	Peter Bosselmann	Representations of Places: reality and realism in city design	Time perception may vary through the form of the environment. Because movements can be read through the (motion) process
14	1999	Suwa, Gero and Purcell	Unexpected discoveries and S-Invention of design requirements: A key to creative designs	The quick and simultaneous turn between the stages of cognition (analysis, synthesis and evaluation)
15	2001	William Pena, Steven Parshall	Problem seeking: An architectural programming primer	• Programming in architecture is the analytic and problem-seeking stage • Design development is the synthetic and solution-seeking stage

Source Author

Problem-solving and decision-making approaches, divergence and convergence, inductive and deductive, analysis and synthesis, bottom-up and top-down, macro and micro

Chapter 3
Normative and Positive Theories

Some design theories judge and evaluate design works such as architecture and reveal some norms or values about the environment. Normative or commendatory views deal with the dos and don'ts; therefore, take certain standards into consideration. These theories are usually coded into declarations, charters, principles and rules.

On the other hand, another group of theories, unlike normative views, just describe what lies there. Such views are called *positive or classificatory views*.

> Normative paradigms tend to recommend and prescribe standards proper for the work, mentioning some dos and don'ts and leading to the desirability and improvement of the existing atmosphere. An example could be when Lynch prescribes in Good City Form (1981) the five criteria of a good city including vitality (sustenance, safety, consonance), sense (identity, structure, congruence, transparency, legibility, unfoldingness, significance), fit (between behavior and place- spatial patterns), access (diverse, equal, local, controlled access to services, material sources, natural environment, information), control (congruence, flexibility, responsibility, certainty) and two meta-criteria of efficiency and justice(Rezaei 2003a, p. 63)

In architecture, for instance, the judgment criteria, according to Vitruvius, include order, arrangement, eurhythmy, symmetry, propriety and economy. Order means architecture should not only have visual sense, but also the various spaces in the plan should serve their respective purposes and missions well. Vitruvius incorporates order with modulation. When considering the efficiency of the plan function, arrangement overlaps order, but it is more in relation with the beauty of the plan composition and its eurhythmy with the façade and volume. Arrangement is a factor that transforms the building into a work of art. Eurhythmy and symmetry are criteria that are connected to beauty. Eurhythmy is the right relationship, proportional as well as formal, of an individual part such as a column. On the other hand, symmetry is the right relationship of all the individual parts to the whole composition. Propriety is how accurately the design corresponds to the common function of a certain type of building. Finally, economy is a quality that Vitruvius defines as the appropriate management of materials and sites with respect to cost and good judgment.

M. Rezaei, *Reviewing Design Process Theories*,
SpringerBriefs in Architectural Design and Technology,
https://doi.org/10.1007/978-3-030-61916-9_3

Alberti[1] considers beauty mixed with function as the full-scale factor in judging a created work of design and subsumes all the criteria of Vitruvius under this factor. Therefore, he poses some questions for the judgment of a work: How well was it conceived (choice, distribution, arrangement, etc.)? How well was it executed (laying, joining, cutting, trimming, polishing, etc.)? How well are the qualities determined by nature (substantive qualities such as heaviness, lightness, density, purity and durability)? How does all this add up (the integration of all factors)? What are the benchmarks of excellence (number, outline, position)? Likewise, John Ruskin, in The Seven Lamps of Architecture, proposes seven criteria of sacrifice, truth, power, beauty, life, memory and obedience. Like Alberti, he considers beauty the main axis of evaluation. In the early nineteenth century, Augustus Pugin[2] published his illustrated pamphlet titled *Contrasts* in which he juxtaposed pictures of industrial cities and their corresponding scenes belonging to the fifteenth century. In this pamphlet, he reproved the features of modern spaces, yet praised medieval spaces. Also in *True Principles of Pointed or Christian Architecture,* which was the summary of his previous lectures on medieval architecture, he started to praise and give credit to this architecture (Hearn 2003, pp. 12, 40–50, 373).

In designing, functionalism has been a normative concept which has reminded designers how they should design. Viollet-Le-Duc's rationalist doctrine held that no building without rationally designed structure is beautiful. Besides, rationalist designing of structures does not necessarily make them beautiful. Based on this doctrine, Louis Sullivan issued his famous motto: 'form follows function'. By publishing an article, titled 'The Tall Building Artistically Considered', in Lippincott's Magazine (57th edition, 1896), he theorized about artistically designing skyscrapers (223–226).

Adolf Loos put his normative paper 'Ornament and Crime' up for debate in the early twentieth century, in the midst of economic concerns, political opposition and ultimately aesthetic criticisms. To Loos, making ornamentation increases construction costs and proportionately decreases the wages of construction workers. Also, from the ideological point of view, ornaments are aspects of the material world and traditionally impose qualitative distinction to objects, which is considered an unnecessary transition from old cultures. Finally, from the aesthetic point of view, he is of the opinion that ornaments, similar to mask, conceal defects of workmanship, materials or basic design (p 279). Paul Scheerbart expresses the advantages of glass and other glazed materials in Glasarchitektur. He analyzes the psychological, sociological and aesthetic advantages of structures sheated in transparent or translucent material (p 16). Le Corbusier and Philip Johnson have explained new and global standards for designing; the former by mentioning his modules in modern designing and the latter, once in aligning modern architecture with international style of design and at another time, react to modern architecture in his book titled *Deconstructivist Architecture.* Peter Eisenman has also questioned accepted design norms with his publications, making effort to create a new viewpoint.

---

[1] Leon-Batista Alberti.

[2] Augustus Welby Northmore Pugin.

In the field of urban design and planning, examples of normative theories are also available. Utopian theories ranging from Surya Mata's linear city to Tony Garnier's industrial city, as well as Le Corbusier's and Frank Lloyd Wright's urban views that gain popularity after the industrial revolution, were considered types of normative views in urban doctrines.

Four years after the publication of Kevin Lynch's *'Good City Form'*, Ian Bentley et al.'s valuable book titled *'Responsive Environments'* (1985) was published, which offered seven key attributes for making a place desirable. These attributes investigate designing a location with people's right to choose and consider permeability, variety, legibility, robustiousness, appropriateness, richness and personalization as desirable standards of the environment. Of course, studies conducted by Bentley's group at this stage generally include built environment and status quo, review urban problems and do not make any references to the natural environment. In 1990, of course Bentley puts forward environmental concerns which were left out by him and his colleagues in 'responsive environments'.

To the same end, Christopher Alexander's seven overriding rules of growth can remind as procedural norms. They include piecemeal growth, the growth of larger wholes, visions, the basic rule of positive urban space, layout of larger buildings, construction and formation of centers (Alexander et al. 1987).

Michael Sorkin, from New York University, also introduces eleven instructions for the design of a good city in his urban design workshops:

1. Reinforce neighborhoods,
2. Make it sustainable,
3. Add green,
4. Secure the edge,
5. Make public places,
6. Be sure rooms have views,
7. Finesse the mix,
8. Elaborate movement,
9. Localize architecture,
10. Defend privacy,
11. Make it beautiful (Sorkin 2001, pp. 5–9).

Apart from these normative theories, all of which have normative implications and seek a utopian concept, there is the anti-normative opposition such as the views of Rem Koolhaas and Coop Himmelblau's group. They are wearing utopian's dress inside out and act as antithesis to the abovementioned theories. Koolhaas announces the death of urban development and deems its existence or nonexistence as the same. He rebels against identifying attributes of the city and apparently stays loyal to architecture. He quotes 'undecidability' from Derrida,[3] 'hyperreality' from Baudrillard[4]

[3]Jacques Derrida.

[4]Jean Baudrillard.

and 'dromology' from Virilio[5] and questions the necessity (and possibility) of considering the supposed norms of the mentioned scholars and experts. Between the thesis of virtues and the antithesis of vices, there also exists a synthesis: Unlike the mentioned schools of thoughts, which possess some sort normative implication, there are theories that are value-free or neutral. When Bill Hillier analyzes public networks and emphasizes their role in drawing up new movement patterns or when Krier Brothers and Gordon Cullen analyze urban anatomy and urban environment, they do draw much attention to normative concepts and implications (Rezaei 2003a, p. 64; 2014a, p. 51) (Table 3.1).

Table 3.1 Normative (value-based) and positive (category-based) design theories

No.	Date	Thinkers	Texts	Ideas
1	1414	Marcus Vitruvius Pollio	The ten books on architecture	Apart from beauty, structure and function, judgment criteria include order, arrangement, eurhythmy, symmetry, propriety and economy
2	1450 (presenting to Pope Nicolai V) 1485 (Published)	Leon Battista Alberti	On the art of building	Beauty mixed with function is the full-scale factor for judging an architectural work
3	1849 1853–1851	John Ruskin	The Seven lamps of architecture stones of venice	The seven criteria of sacrifice, truth, power, beauty, life, memory and obedience for architecture
4	1836 1841	Augustus Welby Northmore Pugin	Contrasts True Principles of Pointed or Christian Architecture	Medieval architecture; the key solution to modern design problems Effort to replace classic temples with Gothic churches
5	1873	Eugène-Emmanuel Viollet-le-Duc	Histoire d'unemaison	No building without rationally designed structure is beautiful. Plus, rationally designed structures are not necessarily beautiful

(continued)

[5]Paul Virilio.

Table 3.1 (continued)

No.	Date	Thinkers	Texts	Ideas
6	1896	Louis Sullivan	The tall building artistically considered	Artistic composition of skyscrapers
7	1908	Adolf Loos	Ornament and crime	Mooney wise, the idea of the aesthetics of form would not need decorative details
8	1914	Paul Scheerbart	Glass architecture	Colored glass destroys all hatred at last
9	1917	Tony Garnier	Unecitéindustrielle	The modern utopia is functionally established based on the centerpiece of industry
10	1948 1955	Le Corbusier	Modular Modular2	Design method based on new ideals and standards
11	1932	Philip Johnson	The international style	Introduction of a new style with certain standards around the world
12	1965	Walter Gropius	The new architecture and the Bauhaus	A structure should be light (i.e., skeletal), simple (i.e., minimal) and straightforward (i.e., explicit)
13	1981	Kevin Lynch	Good city form	Cosmological theory; machine theory; organic theory; the five criteria of a good city including *vitality* (sustenance, safety, consonance), sense (identity, structure, meaning, transparency, legibility, congruence), fit (between behavior and form), *access* (diversity, equity, control), being *well-controlled* (adaptation, flexibility, responsibility, certitude) and two meta-criteria of *efficiency* and *justice*

(continued)

Table 3.1 (continued)

No.	Date	Thinkers	Texts	Ideas
14	1992	Francis Tibbalds	Making people-friendly towns: Improving the public environment in towns and cities	10 golden rules: Places matter most, learn the lessons of the past, encourage the mixing of uses and activities, design on a human scale, encourage pedestrian freedom, provide access for all, build legible environments, build lasting environments, control change, contribute to the greater whole
15	1994	The Politics of Urban Design	Sue Mc Glynn and Paul Murrain	Design is essentially a political mater. Values become an implicit part of ideological baggage acquired in the course of training

Chapter 4
Design Scopes

'Design' is the common ground among fine and applied arts. Apart from architecture and urban design, which are categorized as 'applied arts', most 'visual arts' such as painting, graphic, collage and sculpture contain design. The design scope might be considered from different aspects, for example, from the aspect of design-related professionals, academic disciplines through theoretical or practical approaches, individual or group views and other similar dimensions.

For instance, according to Vitruvius, different professions are associated with architectural design, and taking them into consideration is mandatory in learning or teaching architecture: 'drawing' for making a sketch; 'geometry' for designing and considering the proportions; 'optics' for recognizing the quality of light in the building; 'arithmetic' for calculating the costs and dimensions; 'history' for expressing symbolic features to the client; 'philosophy' for providing the basis for cultivating personal virtues; 'physics' for understanding the laws of nature; 'music' for acquiring mathematical theories (related to acoustic) and to tune weapons; 'medicine' for judging the health conditions of sites; 'law' for informing the rules and regulations related to buildings; and 'astronomy' for understanding the harmony of the universe (Hearn 2003: 31).

Of course, the scope of architecture, as one of the most important design professions, has undergone changes at least in three dimensions. First, in offering new services, for example, in the view of Doxiadis, who reasoned because of social structural changes, this profession has shifted away from merely providing services to kings and aristocrats and has moved toward service providing to citizens of all social classes. Moreover, new architecture-related professions were created such as architectural programming, project management, landscape architecture and new tools were used in the design process such as computers. To Gutman, this has given rise to new clients, especially in façade and interior designing. Second, the scope of architecture has expanded as a result of the rise of a variety of activities. This larger scale is due to the arrival of most of these related professionals in this arena, which made architecture a mix of art and science. Therefore, to perform better architectural

M. Rezaei, *Reviewing Design Process Theories*,
SpringerBriefs in Architectural Design and Technology,
https://doi.org/10.1007/978-3-030-61916-9_4

designing, architects need to have achieved a variety of capabilities. The third point is about the unbalanced competition between architecture and other professions. Apart from the competition between engineering and architecture, proposed by Sweet in 1985, Gutman points out three other professionals in 1988: (1) visual artists especially interior designing, (2) landscape architects and (3) other professionals, who do not possess design skills but have knowledge and experience in dealing with other aspects of the building process, such as building construction managers, contractors and investors (Salama 1995, p 21).

Thus, the concepts of architecture and designing have a broad scope including a wide range of form and space design. Although one can distinguish between the scopes of designing, engineering and sciences, there are collaborations and interactions among them all. Herbert Simon, in his book titled *The Sciences of the Artificial*, believes that designing is the basis of all professions. Besides the fact that designing has different scales, engineering and sciences have always made use of designing to advance their affairs. Designing might be employed in a single-building or multiple-building scale. Urban designing at the scale of a city and industrial designing at the level of designing consumer products require different scopes of their own. Previously, there had never been a distinction between these scopes. And an architect would have given professional opinion about them all. The below diagram depicts how sciences, engineering and designing are related to one another and what their differences and similarities are.

Based on this diagram, it can be explained that the results of engineers' and scientists' efforts are based on pure science and are presented regardless of personal views (above line one in Diagram 4.1), while designing is done based on implicit knowledge, and with a personal view and sometimes even intuition (below line 1). Designers and engineers intend to change the future of the world for the better than the present time, but scientists make effort in understanding the current world (at both sides of line 2). The activities of scientist and designers are human-based, but the activities of engineers are based on technology (both sides of line 3). Farshid Moussavi believes that in the current era, a new opposition has arisen between architects and engineers: Architects attribute unique sensational qualities to the form, while engineers and technicians consider the technical aspects (Moussavi and Lopez 2009).

In spite of this, designing has penetrated in sciences and engineering. The shared points and shared direction of these three have caused processes and design methods to be reviewed in different lights. Tomas Maldonado in Ulm School of Design in Germany[1] is among those who have made efforts to create a balance between science and designing as well as between theory and practice. He and his colleagues, who are the founders of Ulmy Pattern, have established applied anthropology in this school so that its designers are coherently accustomed to society and act like nerve centers in modern and industrial civilizations (Maldonado and Bonsiepe 1964). Horst Rittle, design theorist and expert in design methods and theories, has shown us that activities such as programming, policymaking and engineering have elements of design.

[1] Hochschule für Gestaltung.

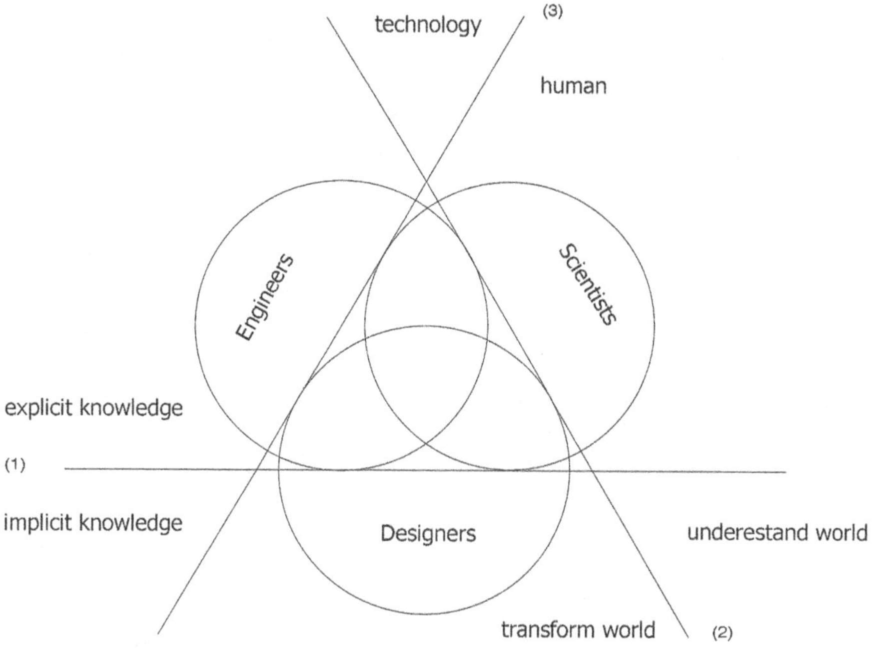

Diagram 4.1 Similarities and differences between designing, engineering and sciences *Source* (Bartneck and Rauterberg 2007, p 738), redrawn by the author

These activities, like designing, are involved in tackling 'wicked problems', while sciences deal with 'tame problems'. Since remembering, exploring and studying all the variables of a wicked problem is not possible for an individual, tackling wicked problems requires lots of people with a wide scope of different sciences as well as engineering and designing fields (Rittle and Webber 1973). About the scope of design and its relationship with sciences and engineering, Broadbent writes:

> Until the beginning of 60s, majors such as system engineering, ergonomics, operation research, information theory and cybernetics, if we do not make mention of new mathematics and calculus, all have been at design theorist's disposal in a very advanced way and numerous events have recognized these fields as the origins of 'design method,' a new and distinctive field of its own. (Broadbent 1973, p 252)

As for the general trend of design process models and creation of design product and problem-solving in the fields of architecture, engineering, environment designing, industrial designing, mechanical and electronic designing, management, computer engineering, psychotherapy and commerce, numerous theories have been offered through evaluating their stages and their methods in a linear, parallel, segmental, cyclic, circular or spiral fashion (Dubberly 2004, p 26).

After being set aside during the 1980s, the issue of methodizing the design process and estimating its problem-solving approaches has once again came to the fore, especially in engineering and industrial design fields. As of this time, as already

Table 4.1 Scope of studies on design process

Year	Architecture	Industrial design	Mechanical engineering	Electronic engineering	Software design	Various/Other
1970	Eastman					
1973	Foz					
1978	Akin					
1979	Lawson					Thomas
1981					Jeffries	
1983	Schon					
1985					Adelson	
1987		Ballay	Stauffer			
1988	Eckersley		Ullman			
1989			Radcliffe			
1990	Chan		Tang Visser		Guindon	
1991	Goldschmidt		Purcell Jansson	Colgan		
1992		Christiaans	Ehrlenspiel		Davies olson	Geol
1993			Fricke			
1994			Lloyd	Ball		

Source (Cross 2001, p 2)

mentioned, sciences, engineering and designing once again have come closer to one another and design aspects flourished one more time in sciences and engineering. During these years, many scientists researched the different fields of design process such as functional tools, management methods, structure of design problem, designing activities and design method philosophy (Vries et al. 1993, pp 16–23).

In 1983, Donald Schön in his book titled '*The Reflective Practitioner: How Professionals Think in Action*' reached the conclusion that the professionals in five fields of engineering, architecture, management, psychotherapy and urban planning, during their problem-solving process, rely more on the design techniques and innovations acquired in practice than on their academic or scholastic learning. According to his study, practical thinking, a kind of thinking that does not occur at schools, is more helpful in design process in all of these fields than mere technical and theoretical rationalism.

By drawing up Table 4.1, Nigel Cross makes an effort in depicting the development process of design activity in various fields of science and engineering. Although this table is not comprehensive and does not show all the studies conducted in this field, it clearly reflects the wide scope and the time-lapse of development for design process in various professions (Cross 2001, p 2).

Designing has been considered as one of the dimensions of planning. Matthew Carmona and John Punter have titled their book 'The Design Dimension of Planning: Theory, content and best practice for design policies'. Some theories of the design scope in the field of urban design are related to measuring the dimensions of the place and its dispersion. Since the interdisciplinary essence of urban planning ranges from architecture to urban geography, paying attention to the place in this profession has covered a wide range of views from international, national and provincial to local, district, neighborhood, square, street, alley and other urban details.

There are scholars who have asked questions about regions, for instance, Walter Christaller by offering the central place theory in 1933, Peter Calthorpe by his 1993 book titled 'The Next American Metropolis' and Kevin Lynch by his 1976 publishing titled 'Managing the Sense of a Region'.

Theories related to the scheme of neighborhoods and neighboring units can be found in the ideas of theorists like Randolph Hester. On the scale of urban spaces (street, square), mention can be made of theories by Robert Krier, Appleyard et al., Mudden and Cooper Marcus. Eventually, from among theories related to details, we can point to Gordon Collin's theories on urban landscape (Golkar 1999, pp 25–26), (Rezaei 2014a, p. 59).

Reiner Benham has offered an exact definition of measuring urban situations within a square mile of periphery. Lynch summarizes different types of design by scales ranging from streets and city squares to urban districts. Christopher Alexander classifies design identifiers from rooms of a home to cities. Most theorists recognize the scale of urban design ranges from architecture (single building) to urban planning (residential settlement).

By expressing the scale of the 'place' in urban design theories, the scale of the 'time' can also be pointed out. Urban designers have often offered their designs as short-term, mid-term and long-term plans. There is usually a direct relationship between the scales of place and time. The smaller the design scope of function with regard to space, then the shorter its time span will become. So less time will be spent for designing a certain site, as compared to designing urban spaces for a certain street. In 1982, Pittas reflects on his observations when writes that the results of urban design compared to those in architecture tend to be more relativistic and less deterministic and at the same time more definite in comparison with planners' perceptions. It is based on the time and scale dimensions that, in his book 'What time is this place?'(1973), Lynch introduces a number of other considerations such as seasonal and diurnal factors along with changing public associations and community history. These are often critically important to human sentiment and values but are usually ignored in designs by designers (Punter 1992, pp 87–88) (Table 4.2).

Table 4.2 Theories based on the scale and scope of design

No.	Date	Thinkers	Texts	Ideas
1	1971	Gordon Cullen	The concise townscape	Urban details
2	1972	Kevin lynch	What time is this place?	Giving importance to time and scale factors (diurnal, seasonal, changing public association and community history)
3	1975	Randolph Hester	Neighborhood space: User needs and design responsibility	Characteristics and fundaments of designing neighborhood units
4	1976	Kevin Lynch	Managing the Sense of a Region	Recognizing the design based on various measures ranging from streets and squares to districts and cities.
				Designing is the base of all professions.
5	1976	Herbert Simon	The sciences of the artificial	Considering a mile as a standard of urbanity.
6	1977	Benham Reiner	Mega structure: Urban futures of the recent past	Design typology from rooms and homes to cities
7	1977	Christopher Alexander	A pattern language	Paying attention to design fundaments in street measures
8	1981	Donald Appleyard	Livable streets	Design principles on the scale of streets
9	1982	Michael Pittas	Education for urban design	Urban design tends to be more relativistic and less deterministic than architecture, but more definite than planners' perceptions.
10	1979	Rob Krier	Urban space	Urban spaces
11	1964	Donald Appleyard	*The View from the Road*	
12	1987	Anne Vernez Moudon	Public streets for public use	
13	1990	Clare Cooper Marcus	People places: Design guidelines for urban open space	

(continued)

Table 4.2 (continued)

No.	Date	Thinkers	Texts	Ideas
14	1994	Stewart Brand	How buildings age: What happens after they are built	The series of six systems, namely site, structure, skin, services, space plan and stuff change at different rates.
15	1997	Varkki George	A procedural explanation for contemporary urban design	Scope of design may range between first-order design (such as architecture) or second-order design(including urban design)
16	2000	Andres Duany, Elizabeth Plater-Zyberk and Jeff Speck	Suburban nation: The rise of sprawl and the decline of the American dream	Inner-city/urban development versus suburban development patterns

Source Author

Chapter 5
Design Participation Theories

The level of the designer's and user's engagement in design processes has always been one of the most significant challenges. With respect to the wide scopes of design, the designer's role can have diverse domains. On the other hand, participatory design is an approach that tries to involve citizens and users in the design process. Historically, in primitive local design or architecture without architects, settlers expanded their eye-catching architecture with local materials and prevalent local construction methods. In his famous book titled '*Architecture without Architects*', Bernard Rudofsky introduces a collection of this kind of architecture.

The participatory level of the beneficiary groups has been a question in architecture ever since, a topic that has been discussed since 1960 in different dimensions of design such as architecture, urban design, urban planning, software design and even graphic design. John Punter believes that the issue of where the power lies in the design process is closely related to the question of who is the urban design client (Punter 1992, p 86). In urban or architectural design, the public role remains controversial. In any case, receiving the demands and budget of the client in designing is not tantamount to participation. Participation in this view is defined as the client's direct engagement in the design process and in reaching the final plan, a point in the requirements and possibilities of which has always been controversial. A group of critics have ruled out the requirement of public participation, considering only the specialist's view or the view of a designer as the criterion of work, while another group has enumerated the complete participation of public groups in design as the cause of success.

Paul Davidoff has posed the theory of advocacy planning. In his view, people will have a more power in participating in urban plans when they seek help from a specialist lawyer. Even if people are not specialists in urban planning, when they

Interactions between designers, users, clients, people, etc. Deterministic or relativistic, pluralistic or individualistic, detailed design or brief design, piece-meal or all-a-piece planning, altruist or egoist approaches.

M. Rezaei, *Reviewing Design Process Theories*,
SpringerBriefs in Architectural Design and Technology,
https://doi.org/10.1007/978-3-030-61916-9_5

are assisted by an urban planner as their advocate, they will be able to hold onto their values. Participation will diminish their sense of anonymity and will boost their sense of trust (Davidoff 1965).

In the sixties, another theory was formulated in the arena of planning called transactive planning. As implied by the name of this view, this theory supports the (face-to-face) interactions with people and the meetings among local community members. As Friedman refers, such a school of planning is a result of the evolution of decentralist planning institutes that support people in imposing stricter controls on social processes (Friedman 1973).

Later on, the supporters of fundamental reforms, whose theories were based on Carl Marx's views, furthered the transactive view to an acceptable extreme. To them, planning has its maximum effect when it is led by unprofessional local planning committees and gives authority to ordinary citizens to experience and resolve their own problems. Susan and Norman Fainstein put planning and policy in a corresponding relationship and make mention of four types of theories: traditional planning, equity planning, democratic planning and incremental planning.

Henry Sanoff's writings have emphasized many instances of participation in design. He defines participation as the face-to-face interaction of individuals who share a number of values important to all. He also believes that our environments are designed by a few individuals, but their design affects a great many of groups. People seldom have a chance to effectively influence or modify the form in which they live. Participation provides the possibility of the presence of users throughout the act of design, i.e., problem definition, the elaboration of the solution and the evaluation of the results (Salama 1995, pp. 33–34, 83).

Participatory design methods prevailed by questioning prescriptive de.signs, especially in the 1970s and 1980s. Prescriptive design holds that the designer acts like a proficient doctor by writing a prescription. Donald Appleyard talks of various types of urban design: The planner sees his model as a totality, from above with a full-scale urban design[1] while the inhabitant sees the present reality from street level. Citizens view the city, according to Appleyard, in different ways depending on their backgrounds, familiarity with the city, patterns of use, educational level and methods of transportation (Appleyard 1976). Roger Trancik has discussed this under the 'overdesigning' idea when writes that 'our tendency is to overdesign and to fill up the space with unnecessary accoutrements' (Trancik 1986, p. 84).

Edmund Bacon believes that a central designed idea will provide the city's backbone and works as the design structure, which can itself become a major creative force to make more meaningful parts. So it can draw the participation of social groups in shaping their environment. This can be done through postponing parts of the design to the implementation phase and urban design regulations. In his analysis of the urban spaces, Bacon recognizes 'democratic feedback' and participatory project review as the creator of a movement in designing that always ends with the design process if the designer is capable of communicating, accepting and restructuring (Bacon 1967).

---

[1] Too much designer's interference.

Table 5.1 Designer's role regarding attitudes and characteristics

The architect role models	Attitudes	Characteristics
Egoist *Elitist-Inactive*	The tendency is to deny or superficially respond to social values, without any involvement in identifying the problem	The architect is paternalistic. His role is to create abstract forms based on subjective feelings
Pragmatist Scientist-Inactive	The tendency is to totally accept the social values as they are without any involvement in identifying the problem	The architect is entrepreneurial. His role is to manipulate and create forms based on the values of
Facilitator Humanist-Interactive	The tendency is to conduct research and then personally respond to social values through the understanding of socio-behavioral contexts. The architect is involved in the process of problem definition	The architect is an interpreter. His role is to manipulate the spaces to accommodate related human activities. It aims at creating a process that enables people to solve their own problems
Technical Assistant Giver Scientist-Reactive	The tendency is to be responsive to the powerless and the poor by being accessible and affordable but without any involvement in identifying community needs	The architect is rationally dealing with physical elements. Other factors that could not be dealt with are ignored
Advocate Bureaucrat-Active	The tendency is to serve the community within their political interests. The architect is involved in the process of problem definition	The architect is political representative. The role is to develop an alternative to a public plan that asserts local interests over the broader public purposes

Source (Salama 1995, p 30)

There have been techniques to improve public participation in design in America and Britain: mandated participations, empowerment and volunteerism, the development of urban design action teams (UDATs), community-based charrettes and comprehensive visioning exercises. Still another public decision-making strategy used by people's place of residence is developing a village design statement (VDS), according to which people are included in developing place design instructions (Punter 1992, p 87) (Table 5.1).

Roger Moore, on a TV program, tried to come up with a collaborative architectural design specifically for private villas while collaborating clients could share their opinions in their deigns over the program, just like the ideas of 'Build Your Homes Yourself' in some magazines. One of the aims of Hassan Fathy in building New Gurena was public participation in constructing and carrying out the village project.

In any case, the level of participation in the scheme can be viewed from another angle. In globalization age, architecture is more committed to considering all members of society with different cultures, needs and the present technology in order to create a lively environment. The movement of architectural method is now making efforts to engage people in the design, like Christopher Alexander, who is looking for designs with people. The approaches that consider behavioral, social and environmental sciences engage more population in the design. A most important new look at participatory design has been mentioned by Nader Khalili. According to him, even with sandbags and barbed wires, there is a possibility of building a shelter by people, based on certain or diverse prototypes.

In the literature of oriental thinkers, there has been emphasis on people's engagement in the design procedure. Salama has reviewed designer's tasks and responsibilities in several texts. In eastern literature, he refers to Touchterman who proposes the architectural designer's role as the intermediary between the responsible authorities and the people. This role might decrease the inaccessibility of local authorities while it may avoid people being considered as marginal groups. Helmy, as Salama reports, has pointed out the change of a designer's role from designing the complete built environment to gradual and developmental designing. The designer is merely a member of a team. Serageldin believes that a designer's role in societies is in transition because designers should define society's sense of its own reality and refine its perception of taste and its cultural expression. Baghdadi deems the dissatisfaction of the traditional elitist or egoist design model to have been caused by the failure of two major design models in responding to society. The one model is an outcome of following the 'international style' which is based on technology as its main focus. The other is the outcome of following a group that perceives architecture as a science of building. They are of the opinion that the architect's role is similar to the role of a coordinator among different professions in the building industry. To him, the new role of a designer, which has more to do with ethical issues and social responsibility, rejects the fact that people can only witness something that they want. It also denies viewing people in a uniform way regardless of history, culture and individual variations. Moreover, the designer's new role is not just acquiring knowledge about the needs of the current users, but also perceiving the needs of prospective users. Eventually, the designer's role does not end in designing or building a product, but it is still effective all through the processes of prediction, adaptation and coordination (Salama 1995, pp. 29–30) (Table 5.2).

Table 5.2 Designer's role and users' role in design (theories based on participation)

No.	Date	Thinkers	Texts	Ideas
1	1909	Daniel Burnham Edward Bennett	The plan of Chicago	Precise and detailed physical design for all parts (full-scale or all-a-piece planning)
2	1964	Bernard Rudofsky	Architecture without architects	Local design done by residents represents a bottom-up model that is anonymous but everlasting architecture
3	1965	Paul Davidoff	Advocacy and pluralism in planning	Planners advocate for the interests of the oppressed and powerless. Participation will diminish grassroots sense of anonymity and will boost their sense of trust
4	1969	James Ackerman	Listening to architecture	The egoist and pragmatist roles of architects; I-give-them-what-I-want or I-give-them-what-they-want model
5	1976	Donald Appleyard	– Planning a Pluralist city: conflicting realities in Ciudad Guayana	Full-scale urban design or all-a-piece planning differs from Pluralist urban design or piece-meal planning
6	1976	Hassan Fathy	Architecture for the poor: An experiment in rural Egypt (An architecture for people: the complete works of Hassan Fathy; 1994)	Building the New Gurena in Egypt with people's participation in construction process
7	1971 1989	Nigel Cross	– Design participation – Designer genes: The nature and nurture of design ability	– Comprising abilities of resolving ill-defined problems, adopting solution focused cognitive strategies, employing abductive or appositional thinking and using non-verbal modeling media – Everybody, in some degree, has the design abilities
8	1978	Strauss and Doyle	The architect as facilitator	The role of the architect as the facilitator; the designer does not solve problems, but creates a process in which people solve their problems themselves

(continued)

Table 5.2 (continued)

No.	Date	Thinkers	Texts	Ideas
9	1978 1979 1988 1990 1992	Henry Sanoff	– Designing with community Participation – Design games – Participatory design in focus – Participatory design: Theory and technique – Integrating planning, evaluation and participation in design	Participation is the face-to-face interaction of individuals who share significant values with everyone Our environments are designed by few individuals, but their design affects a great many of individuals. People are seldom effective on the shape of their place of residence or play a role in reforming it. Participation provides the possibility of the presence of users throughout the act of design, i.e., statement of problem, generalization of solutions and evaluation of results
10	1981	Peter Burgess	Political knowledge and the architectural studio	Architecture's role as the facilitator
11	1983	Jay Garrott	Interpreting value system milieus	The facilitator's role is personal interpretation of the value system
12	1985	Omar Moustafa Baghdadi	The changing role of architect	Designer's role is emphasizing on cultural diversity, providence and a process-oriented view
13	1986	Wolf Touchterman	Training architects: Some comments	Designer's role as the intermediary between the officials in authority and the people
14	1986	John Habraken	Towards a new professional role	Participation is the means of making the user heard and it makes the ordinary social class heard in a serious way
15	1987	Marc François	Community design	Atomistic-holistic design, participation of collective determination, design structure and instructions
16	1986	Roger Trancik	Finding lost space	
17	1987	Christopher Alexander	A new theory of urban design	
18	1967	Edmund Bacon	Design of cities	
19	1991	Ahmed Helmy	The formation of an architect	Change of designer's role from the design of all the built environment to developed design

(continued)

Table 5.2 (continued)

No.	Date	Thinkers	Texts	Ideas
20	1991	Anton (Tony) Nelessen	Aesthetics Control in Europe	Community planning and urban design through public participation using trademarked Visual Preference Survey and Vision Translation Workshop
21	1994	Anthony Costello	Community-based charrettes	Producing community-based charrettes to seek public participation in designs
22	1994	David Lewis	Planning with people	Forming Urban Design Action Teams (UDATs)
23	1983 1986 2011 1994 2001	Nader Khalili	1. Racing alone 2. Sandbag shelter 3. Ceramic houses: How to build your own 4. Sidewalks on the moon 6. Rumi, fountain of fire 7. Rumi, dancing the flame	By providing the basic material to the residents, they can serve themselves as real architects to build their own shelters Proposing the idea of ceramic houses (Geltaftan or earth-and-fire system) and Superadobe system for the housing projects of middle classes and also for the shelters on moon and Mars

Source Author

Chapter 6
Paradigms in Form and Space Creation

A. **Form-Oriented Approaches**

For architects, or any designer, the stage of form-producing process, form-generation process and form-finding or creating space is an inevitable end. However, theories on architectural design process rarely elaborate on the methods of creating the final form in the works of designers. Apart from the discourse that deals with managing the design phases that mainly point to the three stages of analysis, synthesis and evaluation, how do contemporary architects, most of whom do conscious designing, truly create the final product in the midst of the processes of analysis and synthesis as well as form and space? How do they transform the context and content and bring them closer to architectural form and space? In other words, which definite method leads to the creation of their work?

Responding to such questions, some architects have pointed to a certain formal method or design framework in their work, and at times, architectural critics and theorists have extracted indexes from architectural design methods and have attempted to classify them. Some of these classifications have had historical, stylistic and complementary developments while others have been based on the main source of origin. At any rate, such a look at the design process struggles to discover the paradigms and patterns of the methods of creating contemporary architectural form and space. In this view of the design process, at times, more significance is laid on the product than the design process and the emphasis of the cognition stage has changed direction from the inside-to-outside (i.e., from the program and site to the design) to the outside-to-inside direction, which means focusing on the architectural subject and types.

For instance, with a historical approach, Geoffrey Broadbent and Anthony Ward have summarized the architectural design process into four paradigms. The 'pragmatic' method is when the designer makes use of or manipulates the available materials with respect to the climatic considerations to reach the final form with trial and error. The 'iconic' design is when, according to the previous method, the form of the

M. Rezaei, *Reviewing Design Process Theories*,
SpringerBriefs in Architectural Design and Technology,
https://doi.org/10.1007/978-3-030-61916-9_6

building has been reached for a society and is consciously repeated inside or outside that society. Then there is the 'analogical' method, which is formed in different places of residence based on the designers' graphical emphasis and analogical comparisons with available samples, especially by observing familiar forms. Broadbent also points to the 'canonic' method, in which the designer or the designing team gets to create the design work mostly by synthesizing geometric principles. The designer employs a synthesis of aforementioned paradigms, just like Plato created the world elements of fire (tetrahedron), earth (cube or hexahedron), air (octagon) and water (18 dimensional shapes) with volumes consisting of triangles (Broadbent and Ward 1969) and (Broadbent 1973, pp 25–39).

McGinty, based on his own analyses and the comments of other architects, enumerates five concepts in architectural designing: analogies, metaphors, essences, programmatic responses and ideals (McGinty 1979). With a view that is more process-oriented than that of Broadbent, and with regard to the amount of creativity used in the design, John Gero points to three paradigms in designing. The first is 'routine designing', which comes directly from implementing common types and existing samples without changing their variables. Second, there is 'innovative designing', which arises from the status quo, like the previous paradigm, except for the fact that the designer is free to change different variables from the sample. This is like when changing the proportions of a sample leads to the production of new work. In this case, the final design will have a similar structure but a different look because the values of the defined variables are all different. Finally, there is 'creative designing', which employs new variables and creates new types. This last kind of designing will provide the possibility of realizing a paradigm change (Gero 1990). Later on in 2002, Gero recognizes five transformation processes that can lead to new variables in designing: combination, transformation, analogy, emergence and first principles (Gero 2002).

In 'Poetics of Architecture: Theory of Design', Anthony C. Antoniades has recognized 'tangible' and 'intangible' channels for nurturing creativity from which the architectural design is derived. From among tangible channels, he mentions history, historicism and the study of precedents, mimesis and literal interpretation, geometry, materials, the role of nature, art and biography of the architect. To enumerate intangible channels, he mentions metaphor, fantasy, the paradoxical and metaphysical, the primordial and untouched, creating ambiguity, poetry and literature, and the exotic and multicultural (Antoniades 1992).

In his search for 'a new architectural paradigm', and with regard to contemporary developments, Charles Jencks mentions seven recent architectural movements, namely the organitech, the fractal, the blob, the enigmatic signifier, the datascape, the landform and the cosmogenetic (Jencks 2002). Kari Jormakka, Oliver Schürer and Dörte Kuhlmann consider the origin of designing consisting of nature and geometry, music and mathematics, accident and unconscious, rational approaches, precedence, responses to site and generative processes, each of which is, per se, divided into separate parts. They juxtapose classic and common categories with today's classifications (Jormakka and Schürere 2007).

Meanwhile, in addition to the mentioned paradigms throughout history, some have conducted cross-sectional studies within a formal framework and have a single paradigm. They have classified one framework of form and space creation into different categories and have focused on the modes of one method. For example, Karl von Frisch and his son Otto von Frisch and also Mike Hansell studied animal anatomy in advanced architectural creations. Alejandro Bahamon and Patricia Pérez try to complete Frisch's classification of animal architecture based on animal behavior, anatomy and nesting methods and believe in biology and nature as the origin of many contemporary designs (Bahamon and Pérez 2009). Some other, such as Alan Isaac, Roland Knauer, Mark Mückenheim and Juliane Demel, along with the thoughts of Sir Uvedale Price (1747–1829), Richard Payne Knight (1750–1824) and Humphrey Repton (1752–1818), have posed architectural design as having spatial and formal themes, using the form-making alphabet of dots, lines, surface, volume and manner of synthesis and being in relation with factors such as movements, scale, proportion and uniqueness. Based on the works of visual perception psychology, abstract forms have qualities with which one can convey feelings like tension, tranquility, happiness and fear. Therefore, the designer can control users' activities or reactions by applying them in his or her work. By this synthesis, some of them, like the parametric method, point to contemporary approaches in designing. Although the effects of these qualities have been investigated in two-dimensional scale, Mark Mückenheim and Juliane Demel take these factors further, even as far as the contemporary design methods such as the parametric design (Mückenheim and Demel 2012).

Rezaei has studied and summarized the design methodology into different forms of analogical relationships which may vary from very direct (such as pragmatic, programmatic) to more indirect (including metaphorical, enigmatic) analogies (Rezaei 2014a, b) (Table 6.1).

B. **Philosophy and History-Oriented Approaches**

Design paradigms and metatheories might be historically and philosophically coupled with positivism, modernism, postmodernism, or (critical) realism. Various orders or classifications have been argued or made in this field. Giorgio Vasari is among the first people to mention the topic of style in architecture in the sixteenth century. In the contemporary period, Charles Jencks has made one of these classifications based on postmodern versus modern theories. He also mentions the new sciences of complexity—fractal, nonlinear dynamics, the new cosmology and self-organizing systems—as paradigms after claiming the postmodern style (Jencks 1987, p 47).

Based on the same premises, the concept of artistic combination or order has had different paradigms in different periods: composition in classic order, position in modern order and disposition in advanced order. The composition has mostly a hierarchical, cohesive, closed and predictable relationship between parts. Spatial position provides a relativist interpretation of space and of time that links objects freer but still measurable. The contemporary change in paradigms and the new idea of time, as Gausa and his colleagues have mentioned, provide a new informal order upon dispositions, open to individual variation and diversity. If during the classical

Table 6.1 Form-finding and space creation paradigms

No.	Date	Thinkers	Texts	Ideas
1	1958	Gaston Bachelard	La Poétique de l'Espace (The Poetics of Space)	Bachelard implicitly urges designers to base their work on the experiences; therefore, form-finding must be a phenomenological result of personal and emotional response to buildings
2	1969 1973	Geoffrey Broadbent and Anthony Ward Geoffrey Broadbent	Design methods in architecture Design in architecture: architecture and the human	Four complementary design methods in different time periods: pragmatic, iconic (or typical), analogical (or metaphoric), canonic(or combinational)
3	1971	Alan Isaac	An approach to architectural design	Based to visual perception psychology, abstract forms have qualities with which one can convey feelings like tension, tranquility, happiness and fear. (advancing design paradigm according to visual perception of form)
4	1972 1977 1993	Wucius Wong	Principles of two-dimensional design Principles of three dimensional design Principles of form and design	Offering a practical method in the visual grammar
5	1973	Donis A. Dondis	Primer of visual literacy	Common ground among all visual arts includes images, the logical analogy between words and images, analogy of visual elements with the alphabets, the creation of artwork based on this alphabets
6	1979	Tim McGinty	Concept in architecture	Five design concepts: analogical, metaphoric, essential, programmatic and idealistic

(continued)

Table 6.1 (continued)

No.	Date	Thinkers	Texts	Ideas
7	1974	Karl von Frisch, Otto von Frisch	Animal architecture	Possibility of employing animal anatomy in the creation of advanced architecture (advancing a naturalistic paradigm)
8	1975	Francis D. K. Ching	Architecture: From, space and order	Basic form and space alphabet and the principles for organizing them have been depicted graphically through history
9	1980	Bryan Lawson	How designers think: The design process demystified	Adding the narrative design method to Broadbent methods; the three-dimensional model for the constraints and problems of architectural design: (internal, external) (designer, client, user, lawmaker) (radical, applied, formal, iconic)
10	1984	Mikel Hansal	Animal architecture and human behavior	Possibility of employing animal anatomy in advanced architectural creation
11	1990 2002	John Gero	Design prototypes: A knowledge representation scheme for design Computational models of creative designing based on *situated cognition*	Routine, creative, innovative design methods; Five transformational processes that can cause novel creations: combination, transformation, analogy, *emergence*, *first principles*

(continued)

Table 6.1 (continued)

No.	Date	Thinkers	Texts	Ideas
12	1992	Anthony Antoniades	Poetics of architecture: theory of design	Tangible and intangible channels in creativity process: – Tangible: (1) history, historicism, and precedents study, (2) mimesis and literal interpretation, (3) geometry, (4) materials, (5) the role of nature, (6) art, (7) biography of the architect; – Intangible: (1) metaphor, (2) fantasy, (3) the paradoxical and metaphysical, (4) the primordial and untouched, (5) creating ambiguity, (6) poetry and literature, (7) the exotic and multicultural
13	1995	Ashraf Salama	New trends in architectural education; Designing the design studio	Designing is done by the help of induction, deduction, abduction and conjecture-analysis
14	2000	Ernest Sternberg	An integrative theory of urban design	Urban design theories can be classified in four paradigms: 'urban form' (Camillo Sitte), 'legibility' (Kevin Lynch), 'vitality' (Jane Jacobs) and 'meaning' (Christian Norberg-Schulz)
15	2002	Charles Jencks	The new paradigm in architecture	Seven new movements in architecture: organitech, fractal, blob, enigmatic signifier, datascape, landform, cosmogenetic
16	2007	Kari Jormakka, Oliver Schürer, Dörte Kuhlmann	Design methods	The origin of designing consists of nature and geometry, music and mathematics, accident and unconscious, rational approaches, precedent, responses to site, generative processes

(continued)

Table 6.1 (continued)

No.	Date	Thinkers	Texts	Ideas
17	2008	Philip Steadman	The evolution of design: biological analogy in architecture and the applied arts	Various forms of biological analogy consist of organic analogy, natural life taxonomic analogy, anatomic analogy and ecological analogy, which have all been pervasive in architecture and other applied arts ever since early nineteenth century, owing to the advancement of biology
18	2008	Roland Knauer	Basic principles and methodology of design	Major components of form include different combinations of line, surface, body and space, and various forms are created through transformation of these components into one another
19	2009	Alejandro Bahamon, Patricia Pérez	Inspired by nature animals: the building/biology connection	Animals, as a part of nature, can inspire architectural design. (advancing naturalistic architectural design paradigm)
20	2012	Mark Mückenheim	Inspiration: Contemporary design methods in architecture	Dots, lines, surface, texture, structure, color, volumetric transformations, spatial decorations, tracing and superposition, mapping, note taking, folding and parametric designing (advancing the design paradigm based on visual perception of form)
21	2013	Amatal Raof Abdullah, Ismail Bin Said, Dilshan Remaz Ossen	Zaha Hadid's techniques of architectural form-making	(1) abstraction and fragmentation; (2) idea of the ground and gravity; (3) landscaping project and the surrounding context; (4) layering; (5) play of light; (6) seamlessness and fluidity

(continued)

Table 6.1 (continued)

No.	Date	Thinkers	Texts	Ideas
22	2014	Mahmud Rezaei	Design analytica: Reviewing theories and concepts in contemporary design process of form and space	Analogy is the main root of design methodology. It may albeit vary from very direct to very indirect, objective to subjective, internal to external, descriptive to interpretive, realistic to fictional or diverge to converge types of analogy

Source Author

Fig. 6.1 Composition(classical order):Formation, Alive badge of the army, Kelly Field, San Antonio (Texas), 1926 in BOLOGNESI, Kitti, BERNADÓ, Jordi (eds.). Goldbeck, Barcelona, ACTAR Publishers, 1999. *Sources* for figures: Courtesy of Manuel GAUSA

period, the combination of parts, known as *composition*, consisted of their organized synthesis and a subtle change in one part led to a serious change in the whole, it has found a different paradigm during the modern period of combination. During modern period, the combination, known as *position*, takes place as a result of the juxtaposition of parts. In this pattern, the change level of the parts has a weaker relationship with the change level of the whole. In other words, a subtle change in one part delivers change of a lesser degree to the combination, compared to the previous pattern. While in the third paradigm, the combination of the postmodern period, known as *disposition*, is when there are changes in the parts, even in a sudden way, would not bring a noticeable disruption to the synthesis in general (Figs. 6.1, 6.2 and 6.3) (Gausa et al. 2003, p. 175). (GAUSA 2003).

Human's worldview and perspective on the concept of time and space have changed during history thanks to mathematical physics and philosophy. With the help of physics, scientists have been looking for the constants of existence. First, 'weight' was supposed as a constant in physics by Newtonian ideas. Therefore, with a Newtonian view, time and space are absolutes. But considering the connection between weight and place, coefficient of place or gravity remained a variable and mass has been accepted as a physical constant. Einstein mentioned that mass and velocity are related; as a result, he considered time as the constant and the previous hypothesis was rejected. He showed that time is a function of place, so everything is dependent on everything. The law of relativity, like Heisenberg's uncertainty principle, indicated no constant exists. All these affairs revolutionized human's view of

Fig. 6.2 Position (modern order): Alain Resnais, Last year in Marienbad, 1961, in PARKINSON, David, History of films, London: Thames and Hudson, 1996. *Sources* for figures: Courtesy of Manuel GAUSA

Fig. 6.3 Disposition (advanced order):Fence (Photography by Manuel Gausa de Mas Sr., 1960) *Sources* for figures: Courtesy of Manuel GAUSA

the world through the ages. The 'time and space' theory during the classical period turned into 'time-space' theory in the modern period and 'time-space-information' theory in the contemporary period. Table eight depicts how these changes have been in different fields of fundamental design concepts. Also, from the philosophical aspect in the contemporary period, theories and concepts like 'the complexity sciences' have been used as excuses to create an upheaval in design paradigm. The ideas of Leibniz, Jacque Derrida, Gilles Deleuze, Felix Guattari and others have been influential in the formation, development and change of form and space creation method. Theories such as folding, fractal, chaos and the nonlinear system have been among the issues that architects have tried to reflect in their designing.

Peter Eisenman is an architect who made a lot of effort in this field. He mentioned the mathematical differential ideas of Leibniz in the Cartesian mathematical-geometric paradigm. He writes that Leibniz turned his back on Cartesian rationalism in the field of efficient space and discussed this issue that in the sinuous space of continuity, the smallest part is not a dot, but a folding. If this idea is brought to architecture, it will provoke a discussion as below: From the traditional view, architecture is understood as a Cartesian space, which is a set of dot networks. In mathematical studies of change, the idea of the object has changed. For Deleuze, the new object is not providing a framework for the space anymore, but it is rather a temporary modulation that employs a continued change of matter. Constant change takes place through the factor of folding. There is no need for the feature of an object to be determined by one form. He calls this object from an object event (Eisenman 1993).

In his book titled 'Ideas that Formed Buildings', Millard Fillmore Hearn divides fundamental concepts of architectural theories into four paradigms: underpinnings, conventions, principles and convolutions. The first paradigm has been shared throughout history and incorporates the global underpinnings related to all theories. The second paradigm spans from early days to 1800, constituting ritual and conventional theories. This view includes issues like ideal and classical design method, order and proportion. 'Principles' is the title he assigns to mainstream theories ruling between 1800 and 1965. 'Principles' include views such as rationalistic design method, production planning as the principle of designing, true structure as the design framework, honesty in tools (use of materials), decorations and design coherence, restoration and urban design. Since 1965, convolutions theories have been put forward, which have given a new orientation to the design method. Of course, new and old views have been combined and have yielded different modes (Hearn 2003, p 335) (Fig. 6.4).

ORDER FORMAL
UNIVERSAL
LINEAR
DETERMINIST
CLOSED

SPACE AND TIME	SPACE-TIME	SPACE-TIME-INFORMATION
CLASSIC	MODERN	CONTEMPORARY
<AB>: ABSOLUTE	**<RE>: RELATIVE**	**<IN>: INTERACTIVE** (SIMULTANEOUS AND MULTIPLE)

FIXED	*STABLE*	*DYNAMIC*
META-PHYSICAL AND PHYSICAL	PHYSICAL-REAL	REAL-VIRTUAL
ESSENCE (ESSENTIAL)	**MATTER (MATTERIAL)**	**INFORMATION (INFORMATIONAL)**
SINGLE	DIVIDED	DIVERS
ANALOGICAL	MECHANICAL	DIGITAL
RITUAL	*FUNCTIONAL*	*OPERATIVE*
SYMBOLIC	DOGMATIC	OPPORTUNIST (CONTINGENT)
HARMONIOUS	AUTONOMOUS	ACCORDED
EVOCATIVE	ABSENT	REACTIVE
HIERARCHICAL RELATIONSHIP	**POSITIONAL RELATIONSHIP**	**TACTICAL RELATIONSHIP**
CONTINUOUS	DISCONTINUOUS	INTERMITTENT
COMPACT	*FRAGMENTED*	*FRACTAL*
UNIFORM	VARIABLE	EVOLUTIONAL
EXACT	*PRECISE*	*COMBINATORY*
PREDICTABLE	MESURABLE	DIFFERENTIAL
NORM	**TYPE**	**GENE**
PROTO-LOGICAL	TYPO-LOGICAL	TOPO-LOGICAL
FORMAL	ABSTRACT	MIXED
FIGURATIVE	*STRUCTURAL*	*INFRASTRUCTURAL*
SOLEMN	SEVERE	EASYGOING
CEREMONIAL	STRICT	UNINHIBITED
PURE	PURIST	CROSSBRED

CODE	**RELATIONSHIP**	**COMBINATION**
CONTROL	**ORDER**	**SYNERGY**
FLAT (2D)	**VOLUME (3D)**	**LANDSCAPE (4D)**
COMPOSITION	**POSITION**	**DISPOSITION**
(REGULATION)	**(CORRELATION)**	**(DECISION-COMBINATION)**

NEW MECHANISMS:
BEYOND LINKS- EXTROVERTED FORMS

<IN>FORMAL ORDER
MULTIPLE
NON-LINEAR
IN-DETERMINIST
OPEN

SPACE-TIME -INFORMATION AND NEW ARCHITECTURE

<IN> PROPERTIES	
<IN> DETERMINATION	<IN> FINITENESS
<IN> STABILITY	<IN> TERACTION
<IN> COHERENCE	<IN> DISCIPLINE
<IN> COMPLETION	<IN> FORMALITY

Fig. 6.4 Changing concept of space since the classical times until the contemporary period. *Sources* Courtesy of Manuel Gausa; GAUSA, Manuel: Comparative table and variated definitions of 'Space-Time-Information' published in *Quaderns* n. 222, 1999, in *Open* (Barcelona, Actar 2010) and in Gausa et al (2003), *The Metapolis Dictionary of Advanced Architecture*, Barcelona, Actar, p. 626 (Gausa et al. 2003, p. 626)

Chapter 7
Efficient Tools and Sources in Design Process

A. **Representational Approaches**

Representational tools for architectural or design works have made great progress so far. Words, phrases, architectural drawings, freehand sketches, photography images, model making and computerized 3D tools have all been at the service of visualization and representation of architectural space.

It has been controversial whether generators of architectural design have been reliant on visualization tools. Bruno Zevi has pointed to the dependence of architectural space complexity to the capacities of tools for visualizing such spaces. Using architectural dialogues, freehand sketches, diagrams, plans, cross-sections, 3D perspectives and computerized facilities are among designing guide tools that are influential in the design process or product.

As Broadbent has discussed (1973, pp. 343–353), making a personal checklist of photos or phrases is among the idea generation methods. One of the items of the checklist can be Jung psychological tests in which the individuals should mention anything that comes to their minds. Osborn (1963), Gregory (1963), Matousek (1963), Eder and Gosling are among those who have pointed to this method. Interactional techniques include linear diagrams (Guerra 1969), Norris (1962) diagram or table based on Fritz Zwicky's (1948) view or psychoanalytical techniques such as brainstorming (Osborn) and synectics (Gordon 1961; Alexander 1965).

Adel Tewfik and Khalil Foz have studied designers' behavior in design sketches and etudes. To them, in their sketching, designers make use of four sources to create form: program, consultation with the clients and experts, personal knowledge and abilities as well as representational media. From their observations, Tewfik and Foz have concluded that more adept designers, besides having access to more resources, possess greater ability in conforming these resources to human's mental restrictions, and they also employ more 3D representational methods in their works compared to others (Tewfik and Foz 1972). They believe 'product'representational tools work effectively in the 'process' of design, anyhow.

M. Rezaei, *Reviewing Design Process Theories*,
SpringerBriefs in Architectural Design and Technology,
https://doi.org/10.1007/978-3-030-61916-9_7

Fig. 7.1 Organization drawings related to program studies. *Source* (Gross and Do 1995, p 7), redrawn by the author

In '*The Dialectics of Sketching*', Goldschmidt points to a method for justifying designing in a reasonable way. Goldschmidt calls this method interactive imagery, in which sketches accompany design moves and controversies, and suggests that the difference between seeing as a formal discussion and seeing as a non-formal discussion enables the designer to bridge the gap between formal and conceptual (non-formal) designs (Goldschmidt 1991).

After interviewing several designers, Ellen Yi-Luen Do classified design processes along with their relevant sketches into three categories: organization, conception and fabrication. Therefore, three drawing tools are recognized; organization drawings illustrate functions, activities and the relations between parts; in conception drawings, the designer sketches to discover ideas and to apply forms and metaphors. In this type of sketching, designer's goal is more or less reaching the formal and artistic solution. As a result, geometry and geometric principles and esthetics of form are given attention like rhythm and harmony; in the final stages of design, especially when the designer gets close to implementation phase, the designer's third category of drawing and visual activities comes closer to reviewing methods of fabrication, structure and utilities. In this type of technical sketches, details and networks are more helpful than those of other sketches (Gross and Do 1995) (Figs. 7.1, 7.2 and 7.3).

Goel believes the design process to be consisting of lateral transformations (nature of ideas) and vertical transformation (idea development or details), which leads to the process of final form creation. According to the research, the amount and quality of sketching are helpful in the improvement of lateral transformation, i.e., idea changing. These transformations, per se, are very important in advancing the initial process of

Fig. 7.2 Conception drawings to review figure and form. *Source* (Gross and Do 1995, p 7), redrawn by the author

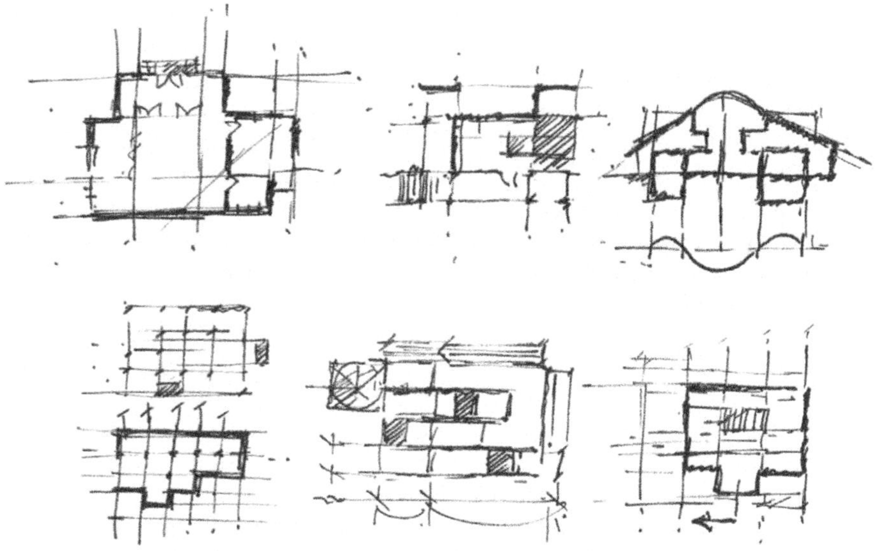

Fig. 7.3 Drawings with more precision. *Source* (Gross and Do 1995, p 8), redrawn by the author

architectural design. It is also accepted that the sketch in architectural design not only acts as an external memory, but also provides visual clues and creates a physical base that forms design ideas(Suwa et al. 1998) (Hay et al. 2017).

According to Suwa and Tversky's studies, who observed planned experiences for a group of designs by architects and students, design tools providing external representation of design (such as freehand sketches, diagrams, prism, graphs and notes) are necessary to clarify design ideas in the early steps of the design processes (Suwa and Tversky 1997).

Computer has also been used as an architectural design tool by many professionals. Some have taken it as a threat for the future of their profession, and some others have considered it a magical tool in designing. Members of the third group, like Frank Raymond Leavis, have rejected its use altogether in enriching human life and believe it to be a cause for the decline of human beings. Broadbent regards it necessary to correctly guide and carefully balance what the computer is capable of doing and what it is not. In his opinion, if we design new computers that are capable of working with the human brain more efficiently, then these computers will provide humans, more than any time, but also the freedom of using intelligence. As Lady Welby puts it, it will carry out anything we know in any way that we order it, and thus, it is up to us to decide what this tool will be doing (Broadbent 1973, p 299).

Ellen Yi-Luen Do believes that designers have used computer as a tool to support the tripartite activities of organization, conception and fabrication sketching, as mentioned above. For instance, functional analysis, spatial layout and knowledge-based evaluation programs support organizational decision-making, or the CAD software, by having a databank of the parts of a building, supports fabrication designs. Computer support has so far been scarce for conception designs (Gross and Do 1995).

Gero acknowledges drawing and designing to be two important computer functions in the design process. To him, computer tools both help represent the design topic and cause the formation of new design compositions (Gero 1994).

Digital morphogenesis is a term in architecture that respects more than employing the digital media as a representational tool, but it employs them in the direction of creating form and conforming it to the context or responding to the problem of designing. What Roudavski has stated holds the analogical or metaphoric communicational concept with morphogenetic processes of nature such as gradual growth or adaptation and concepts like emergence, self-organization and formation (Roudavski 2009).

Moreover, Neil Leach points to computer-assisted emergence design, in which modeling is done from the design logic, instead of coming from a topic. He has analyzed the used computer as a tool in creating form and space in architecture and urban construction (Leach 2009). He mentions the idea of the 'swarm city' by reviewing the speculations of Steven Berlin Johnson in *'Emergence: The connected Lives of Ants, Brains, Cities, and Software'*, and the reflections of Tom Verebe in London's Architectural Association (AA) and the works of Hernan Diaz Alonso and François Roche, one of them being formed based on a sci-fi on the future of Los Angeles, pose the idea of 'swarm city'.

Based on a lecture series at Pennsylvania University in 2002 titled '*Architecture in the Digital Age: Design and Manufacturing*', Branko Kolarevic thinks of the digital era as bringing about the emergence of a distinctive type of architecture. He also thinks that the computer has provided a good opportunity to redefine the architect's role in the building process and has brought the design, fabrication and parts and also the architect and civil engineer and the executor of the building project closer to each other throughout the design process (Kolarevic 2003).

Digital designing or architecture is the title that is given to a range of computerized techniques in classifying design methods. Using computer software is not necessarily tantamount to digital designing. Most current software programs are used for the purpose of architectural drawing or plan design, which is not a fundamental stage of form and space emergence. Rather, using the inherent characteristics of computers in the design process means disintegrating the process into discrete patterns and aggregating the patterns into processable templates by using a computer in order to create space. The forms produced by digital design are different from those created by the world of manual designing (Rezaei 2014a, p. 82).

In his '*What Designers Know*', Bryan Lawson deals with the skill of exchanging theoretical, experimental and professional design knowledge by means of drawing, interactive and computerized tools. He points to diagrams and pictorial, educational, consultative, experimental, fictional, suggestive and calculative drawings. He also discusses different roles computer plays ranging from architectural drawing and negative force to modeling, critical and networking role (Lawson 2004).

B. **Internal and external sources of design procedures**

Even though Vitruvius's three principles,[1] which base architecture on form, structure and function, are known to everyone, one can recognize cognition components themed in designing inside and outside the project, especially those derived from rational approaches.

Jean-Nicolas-Louis Durand emphasizes the functional program of the building. Sociological function and hierarchy of building types range among the most important factors of design. He believes that at the first degree, it is the complexity or simplicity of the program which shapes the form of the architectural plan and consequently volume. In his design process, practical aspects such as the features of site, structure and materials are relegated to a next level of importance. To Viollet-le-Duc, the architect's work should be a response to the client's view. Therefore, each design should follow the functional program of the building. The second factor after program is the site or the context of the project. Together, program and site make the client a partner in the creation of the form. About the site, le-Duc points to issues such as the kind of soil (clay or sand), the location and depth of bedrock, waste drainage and the availability of freshwater. He also considers legal terms, conditions and restrictions of building construction in the area of town where designing is taking place as the factors influencing design. These terms and conditions include zoning for building types, height of construction, proximity to the street front, density and floor area ratio (FAR). After knowing all these factors, the designer can opine on the optimal location and orientation of the new design. This, per se, depends on factors such as views from the site, sunlight direction, prevailing winds, the appropriate access and approach to the site, the most feasible location of service areas and service access

[1]The Roman architect Vitruvius in his treatise on architecture, De architectura, asserted that there were three principles of good architecture:

- Firmatis (Durability)—It should stand up robustly and remain in good condition.
- Utilitas (Utility)—It should be useful and function well for the people using it.
- Venustatis (Beauty)—It should delight people and raise their spirits.

routes. After meeting all these criteria, it is time to think about the artistic composition of the building or complex of the buildings and exploitation of the site's natural features. At times, studying the site provides so many inspiring or restricting factors that, to le-Duc, those factors shape the design in designer's mind. Spatial functions attained from the program will be arranged with respect to their priority and in relations to the site. Dimensions and sizes, spatial standards, circulation and access areas in the building, entrances, interconnecting doors, fenestration and openings should all be seen in the architectural design before implementation. These are first applied in the plan and then in the sections, elevations and the structural framework. The selection of material and structure is the next step, and it is very cost-sensitive. The openings should be designed with respect to the interior requirements not from the aspect of design form. Decorations can reflect the building structure (Hearn 2003, pp 182–189). Also, when describing the principled method of architectural design, Broadbent has pointed to spatial and environmental requirement and the relationship between them (Broadbent 1973, pp 428).

Aside from the internal factors, which are traceable and applicable in the recognition of architectural design, the criticism of modern rational view by postmodern and poststructuralist views brought forth external components in architectural design. Robert Venturi points to the absence of cultural concepts in modern architecture's design methods. Even though he does not explicitly offer a solution to this issue, his attention to the present cultural and historical types and samples regains importance. Besides classical approaches to architectural types, attention to similar samples in the design process was received both during the Renaissance, to insist on ancient values, and also in criticizing modernism for letting meaning and history in the works. According to the modern view, design method is a problem-solving process that begins with a list of functions. But with a postmodern look, there are other effective issues in the design process, apart from function and the building program. Besides functional requirements for erecting a building, the client has a philosophy and purpose in mind. The cultural role of the building is an important factor that must not be neglected by users, clients, people and the designer. Along with philosophy and culture, topics such as history, meaning, place, city and urban concepts, moral and political agenda, and modern views at physical aspects should also be taken into consideration; various aspects of sciences and arts, too, need to be pointed out, and they have had a very strong role in the design process, especially the postmodern design (Nesbitt 1996, pp. 40–46). While categorizing the constraints of the problem to internal and external ones, Bryan Lawson also attends to their relationship with the designer, client(s), exploiters and lawmakers. His 3D model has also made clear fundamental, functional, formal and symbolic aspects of the design problem (Lawson 1980, pp. 92–106). Chris Abel recognizes the below components as architectural design sources and determines the scientific fields related to each component next to it:

1. *Location: geography, geology, ecology, climatology*
2. *Social form: philosophy, anthropology, sociology, history, economics, politics*
3. *Building type: typology, morphology, planning*

4. *Technology: engineering, building materials and construction*
5. *Cost: economics, quantity surveying, building management*
6. *Aesthetics: philosophy, arts (Abel 1988, p 162).*

Therefore, according to the views of these scholars, the sources of designing form and space can be summarized in Tables 7.1 and 7.2.

Table 7.1 Categorizing external and internal sources of creating form and space

External design sources	Internal design sources
Social issues Cultural aspects Similar projects Sense and meaning History Philosophy Place and context City and urban concepts Moral and political agenda Sciences Arts	Program and function Dimensions and measures Standards, rules and regulations Legal terms and conditions (zoning, height, proximity to the street front, FAR) Hierarchy and circulations Site (soil, bedrock, drainage, freshwater availability, views, sunlight direction, prevailing winds, etc.) Structure Materials Costs Client's opinion

Source Author

Table 7.2 Theories on design sources and representational tools

No.	Date	Thinkers	Texts	Ideas
1	1948	Bruno Zevi	Architecture as space: How to look at architecture	The architectural spaces are highly related to the capacity of representational tools which have created those spaces
2	1957	Alex Osborn	Applied Imagination: Principles and procedures of creative writing	Checklist and brainstorming
3	1961	William Gordon	Synectics: The development of creative capacity	The group activity of using analogy and metaphor with the guide of a leader, not accidentally
4	1963	Richard Langton Gregory	Distortion of visual space as inappropriate constancy scaling	Using a checklist as a creativity technique
5		Robert Matousek	Engineering Design: A systematic approach	

(continued)

Table 7.2 (continued)

No.	Date	Thinkers	Texts	Ideas
6	1965	Tom Alexander	Synectics: Investing by the madness method	Leading the group by asking questions about personal metaphor, direct metaphor, symbolic metaphor for problem-solving
7	1973	Adel Tewfik and Khalil Foz	Some observations on designer behavior in the parti	There are four sources for design sketching: program, consulting experts, knowledge and representational tools
8	1988	Chris Abel	Analogical Models in Architecture and urban design	Main issues in architectural design come from: location, social form, building type, technology, cost, aesthetics
9	1991	Gabriela Goldschmidt	The Dialectics of sketching	Interactive imagery: reasoning and debate method for the design (Design movements and discussions, which can be formal or non-formal)
10	1991	William J. Mitchell Malcolm McCulloch	Digital design media (Architecture)	Role and importance of digital design
11	1993	Ellen Yi-Luen Do	Stages of drawing conception diagrams in design process	Three types of designing and sketching activities during the design process: organization (to show functions and relations), conception (to find the solution and form of the work), fabrication (to review technical and executional details of the design)
12	1995	William J. Mitchell	City of Bits: space, place, and the infobahn	The effect of digital space on the city
13	1995	Vinod Goel	Sketches of though	Horizontal changes (idea changes) and vertical changes (idea development)

(continued)

Table 7.2 (continued)

No.	Date	Thinkers	Texts	Ideas
14	1996	Malcolm McCulloch	Abstracting craft: The practiced digital hand	The role of computer is beyond that of a tool. Traditionally, there is a growing correlation between digital works and professional design works
15	1998	Terry Purcell, John. S Gero Suwa	The role of sketch in the preliminary stage of design process	The sketch in architectural design not only acts as an external memory, but besides providing visual clues, it creates a physical base that forms design ideas
16	1997	Masaki Suwa, Barbara Tversky	What do architects and students perceive in their design sketches? A protocol analysis	External representation of design (such as free-hand sketches, diagrams, prism, graphs and notes) is necessary to clarify design ideas in the preliminary steps of the design process
17	1999	Peter Eisenman	Diagram diaries	Diagram is visualization of something that is not that thing anymore. So it can make something else visible. The diagram is used in two forms of analytical and productive
18	2001	Steven Berlin Johnson	Emergence: The connected lives of ants, brains, cities, and software	What modern sciences can teach us about thoughts, societies and ourselves
19	2004	Bryan Lawson	What designers know	The skill of exchanging theoretical, empirical and professional designing with the help of drawing tools (visual, educational, consultative, empirical, narrative, suggestive, calculative and diagrams), interactive tools and computational tools (in the role of a draftsman, negative force, model maker, critic)
20	2005	Peter Eisenman	Feints	Diagram as the structure of an individual's language for architectural expression

(continued)

Table 7.2 (continued)

No.	Date	Thinkers	Texts	Ideas
21	2005	Branko Kolarevic	Architecture in the digital age: Design and manufacturing	The digital age is the herald of a distinctive architecture and has provided a good opportunity to redefine the architect's role in the building process, and it has brought the design, analysis, construction, execution, and also the architect, engineer and project manager close together
22	2006	Kostas Terzidis	Algorithmic architecture	Architectural design can be simulated and produced based on an algorithm in which the constants and variables are swapped
23	2009	Stanislav Rodovsky	Toward morphogenesis in architecture	Digital morphogenesis is an analogy of natural morphogenetic processes, which takes advantage of gradual growth concepts, adaptation, genesis, self-organization and formation
24	2009	Neil Leach	Digital cities	Swarm intelligence, swarm city

Source Author

Chapter 8
Place and Non-place Theories

Design theories can also be viewed based on place and time. Christian Norberg-Shulz has well-famed writings on space and place. In a short essay titled 'Place' (1979), beautifully and artistically, he based his philosophical discussions of 'place' on the poem, 'A Winter Evening' by George Truckell, which had previously been employed by Heidegger to describe the nature of language, and he viewed on life as it mingled with the aspect of place.

It is especially inspired by Lynch and Collin that the idea of place has been widely put to practice by environmental and urban psychologists and human geography experts who studied the concepts and meanings ascribed to place by people and vice versa (e.g., Relph's and David Counter's efforts in 1976 and 1977). On the basis of different aspects, nowadays sense of a place is completed with three components: (1) physical setting, which includes the outward aspects, formal dimensions and the physical environment (such as landscape, built form and furniture); (2) behavioral aspect, which includes the stream and pattern of pedestrian and motorist movements and activities (also the noises, smells and movement features); and (3) meaning or perceptional aspect, which is comprised of public imaginations and their mental evaluation (such as qualitative evaluations, attractions, legibility, way findings and cultural aspects). As a result, in 1991, like some others, Punter draws the triangle of sense-of-place components with the help of such views. The concept of 'place-making' has been in common practice among architects and urban developers since 1970. This concept is often used for placemaking aimed at attracting people and creating livability and vitality (Rezaei 2003d, 2011) (Marzbani et al. 2020).

There is a famous magazine titled 'Places: Forum of Design' that reviews experiences and outcomes of this issue. Here, too, the three physical, activity and mental components have a lot of influence in placemaking. The triangle of place will find greater dimensions based on the aforementioned three components by some scholars, including Matthew Carmona. For instance, Matthew reviewed it in his octagon of urban design theories: ecological, spatial, morphological, contextual, visual, perceptional, social and functional aspects (Carmona 2001) (Diagrams 8.1

M. Rezaei, *Reviewing Design Process Theories*,
SpringerBriefs in Architectural Design and Technology,
https://doi.org/10.1007/978-3-030-61916-9_8

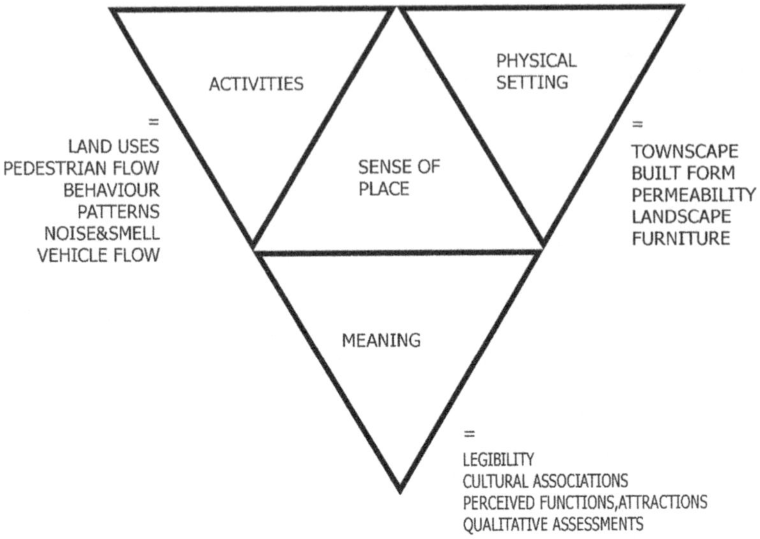

Diagram 8.1 Design considerations and components of a 'sense of place'. *Source* (Punter 1992, p 76), redrawn by the author

and 8.2; Tables 8.1 and 8.2).

Although since the 1960s with Lewis Mumford's (1962) ideas of anticity, Melvin Webber's (1964) non-place urban realm and Doxiadis' (1966) ecumenopolis, the collapse of place and the degradation of its importance in comparison with time had been hypothesized, but it was as of the 1990s, along with the advance of technologies that the scholars and experts began to theorize expansively about the annihilation of urban life and urban construction and minimizing the significance of geographical factors. To Frances Cairncross, hinging on the ideas from Alvin Toffler, digital spatial facilities equaled the death of distance.

To Luigi Prestineza Puglisi, space has three main characteristics and new logic in the electronic age coming right after industrial society which has carried the logic of the 'machine for living'. He explores those three in some architectural projects including *Beaubourg or George Pompidou Center* in Paris, Times Square in New York and Egg of Wind by Toyo Ito in Yokohama:

> Here we enter a new age, marked by electronics and characterized by three phenomena: immateriality, sensoriality, multimediality.
>
> Immateriality is primarily expressed by transparency. Zevi......compares Beaubourg to the description of one of the Italo Calvino's invisible cities: it has no walls or cellings and no floors: it has nothing which makes it look like a city, except for the pipes...Sensoriality is the capacity of a structure to interact with outside world.....Lastley, multimediality represents the choice to transform the building into an organism capable of conveying messages using various media, integrating them into the building fabric. (Puglisi 1999, pp. 7–8)

Breaking the border between the real and virtual spaces or physical and electronic ones is clear in postmodern works and thoughts. This is to the extent that various

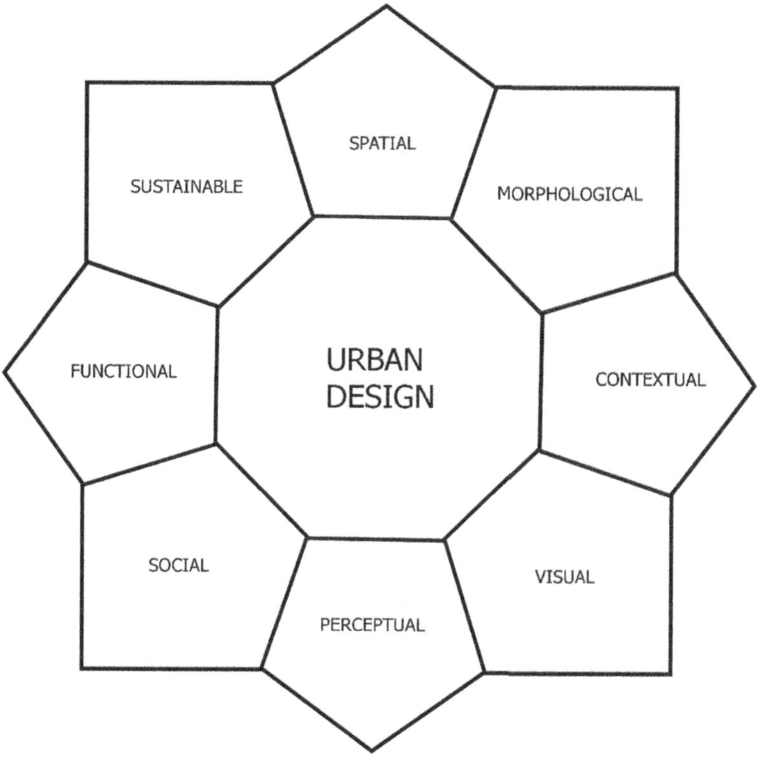

Diagram 8.2 Urban design aspects. *Source* (Carmona 2001, p 282), redrawn by the author

terms have been used for this concept. Marcus Novack talks of *liquid architecture*. Stephen Perla makes mention of *media space*. Hardison draws upon the *horizon of invisibility*, Bill Viola uses *data space*, and William Gibson utilizes *cyberspace*.

To Kathrin Hörschelmann, the concept of territory is nowadays substituted for a sense of spaces and lots of which are rich with media experiences. Cultures are overlapping and are not distinctive anymore. Their boundaries have collapsed and merged.

According to Houghman, these spaces, instead of making dialogues, are after creating feeling. Apart from non-placeness, the other point about these spaces deals more with the loss of identity or originality which are affected by information and communication technologies. What was once considered and functioned as the bedroom area can now be transformed into a global central office. What had only been a garage is now changed to function as an international terminal.[1] Therefore,

[1]However, the bedroom or garage has also been more meaningful spaces in traditional periods which in modern age are transformed into areas with specific 'functions', to sleep or to park a car. In traditional societies, the bedroom is still called with names conveying more cultural meanings.

Table 8.1 Classification of place theories based on Counter and Punter's three components and Carmona's eight components

Aspects of a 'sense of place'		Related theorists
Level one (based on Counter and Punter's model)	Level two (based on Carmona's model)	
Physical	Ecological	Rogers (1997), Coupland (1996), Jenks et al. (1996), Bulvarz (1993), Barton et al. (1994), Harg (1969), Espirn (2000), Hough (1989), Steiner (1995), Bently et al. (1990), Mostafavi (2010), DoE (1990)
	Spatial	Peter Calthorpe (1993), Garreau (1991), Frey (1999), Lynch (1976), Hough (1990)
	Morphological	Sitte (1889), Martin (1972), Krier (1979), Hillier (1996), Whitehand (1981), Alexander (1977, 1987), Rossi (1982), Trancik (1986), Kelbaugh (2002)
	Contextual	Worskett (1969), Tibbalds (1992), Sharp (1946), Nairn (1955)
	Visual	Unwin (1909), Cullen (1961), Tugnutt and Robinson (1987), Buchanan (1988), Gibberd (1953), Mozayeni (1996), Cantacuzino (1994), Moughtin (1995), White (1999)
Conceptual	Perceptual	Lynch (1960), Straus (1961), Bacon (1975), Relph (1976), Norberg-Schulz (1980), Appleyard (1981), Altman, Wellvile and Wohlwill (1985, 1976), Knox(1987), Moore et al. (1985), Jencks (1988), Nasar (1990), Zukin (1995), Sircus (2001)
Behavioral	Social	Jacobs (1961), Gehl (1971), Newman (1972), Sennet (1970, 1972, 1977), Bentley et al. (1985), Whyte (1980, 1988), Oldenburg (1989), Goldberger(1996), Banerjee (2001)

(continued)

Table 8.1 (continued)

Aspects of a 'sense of place'		Related theorists
Level one (based on Counter and Punter's model)	Level two (based on Carmona's model)	
	Functional	DoE and DoT (1992), Stephen Carr, Mark Francis, Leanne Rivlin and Andrew Stone (1992), Mc Cormac (1994), Lang (1994, 1987, 1995), Cooper Marcus and Sarkissan (1986), Rapoport (1977, 1982), Hillier(1996)

Source Author with an adaptation of Carmona (2001), (Rezaei 2013b, 2014a)

new expectations of space have emerged, which have made its function really versatile, its users and residents transitory and its form free from any functional worries. Consequently, space has been liberated from the restraint of usage, form, capacity, user, physical form and even designers.

By getting their hands on information and communication technologies, digital networks and participating in projects, urban designers and architects are capable of studying and analyzing a building from far away before the building is built and without even being present at the site. Isozaki never saw his Team Disney building in Epcot, Florida, directly with his own eyes. But he directed it with the eyes of his mind. It is here when one may say that event spaces, empirical architecture and dynamic programs are substituted for traditional physical ones. What is noticeable about Isozaki's design process is his unique interaction with his building. Korteknie and Stuhlmacher even went beyond this, and by designing Las Palmas in Rotterdam, they issued a statement for flexible, temporary and all-place or non-place construction (Rezaei 2004).

Three fields of studies, at least, must be applied into the body of knowledge of design processes: cyber geography, dealing with the digital networks and the representation of space of flows; cybernetics and systematic planning which contains algorithmic approaches; and the planning of mediated spaces which combines virtual and non-virtual spaces within the public open spaces (Townsend 2003) (Rezaei 2003c, 2004, 2005).

Table 8.2 Some non-place theories

No.	Date	Thinkers	Ideas or texts
1	1962	Lewis Mumford	Anticity
2	1964	Melvin Webber	Non-place urban realm (the society of interests is composed of individuals with similar interests and does not have a certain physical limit)
3	1966	Constantinos Apostolou Doxiadis	Ecumenopolis
4	1960 1967 1973 1975	Peter Wilmot Michael Young	People are inclined toward megacity life styles, i.e., toward separating from their neighbors and their neighborhood centers. There is little compatibility between forms of community and physical complexes. Creating nonlocal communal interaction for residents leads to further urbanization (practically in designing Cumberland in Scotland) Society without place proximity (Milton Keynes)
5	1974	Henri Lefebvre	Hyperspace
6	1984	Fredric Jameson	New hyperspace
7	1985	Kenneth Jackson	Centerless city
8	1988	Lawrence Kubie	Information superhighways
9	1989	Manuel Castells	Information city
10	1989	Mark Hepworth	Electronic network, virtual geographies
11	1991	Saskia Sassen	Global cities
12	1994	Howard Rheingold	Virtual communities
13	1995	William Mitchel	City of bits
14	1996	Christine Boyer	Cybercities
15	1996	Stephen Graham and Simon Marvin	Telecommunication and the city
16	1997	Marcus Novak	Liquid architecture, deconstruction
17	1999	William Dutton	Society on the line
18	2000	James wheeler, Yuko Aoyama, Barney Warf	Cities in telecommunication age
19	2005	Mahmud Rezaei	Urban design in the third millennium (The role of ICT in urban design)
20	2013	Anthony M Townsend	Smart cities: Big data, civic hackers, and the quest for a new Utopia
21	2015	Antoine Picon	Smart cities: A spatialised intelligence

Source Author with an adoption of (Rezaei 2004, p. 98 and 2014a, p. 93)

Conclusion

Design process is being employed in many fields including computer programming, environmental theories, management, landscape architecture and urban design. This process has also various aspects based on two main stages of analysis and synthesis. Considering styles or 'manners of thinking', for example, it can be divided into a spectrum ranging from empiricism to rationalism. Furthermore, intuitive, rational and participatory designing work as three main models of this spectrum.

Another aspect of design methodology may deal with the orientation or 'directions of the process'. It may emphasize on the 'product' by synthesizing several elements and converging them all in the design solutions. Or it could work as an analysis 'process' diverging from the core of the problem. Nowadays, process-oriented views have gained more importance.

This is to the extent that from the aspect of the level of 'designer's role' and people's participation, there is more emphasis on participatory design than individualist and elitist design. Another words, bottom-up way of design thinking differs from top-down processes in designing. While the former engages more design actors negotiating on development alternatives, the latter works as if the designer is the leader and the only form-giver to the process of design.

As for employing the 'design tools and sources', using computers is now considered the most recent dimension of the design process so that it has put forth new paradigms for designing. As a result, in the course of history, form and space 'paradigms or models' have been transformed into a novel design process. Therefore, from another view, if we take paradigms of creating form and space posited by architecture and urban design theorists into consideration, we can measure various dimensions and aspects in design processes. Some designers apply internal sources and adhere to site and program constraints, whereas many architects may use direct or indirect analogies with phenomena out of the design problem.

Even though design theories have provided coherent ideas on place-based concepts, fewer scholars have classified non-place theories. In the last discourse of this book, however, the author has tried to compare place and non-place dimensions

M. Rezaei, *Reviewing Design Process Theories*,
SpringerBriefs in Architectural Design and Technology,
https://doi.org/10.1007/978-3-030-61916-9

Table 1 Depicting the aspects of design process

Row	Aspects of design process	Various cases in the course of history
1	Manner of thinking	From empiricism to rationalism Intuitive, rational and participatory designing
2	Process direction/orientation	Product orientation to process orientation Analysis to synthesis Problem orientation to solution orientation Convergence to divergence
3	Position	Positive to normative
4	Design scope	Science, engineering and technology to various arts
5	Designer's role	Design-centered to participatory outlook Top-down to bottom-up
6	Models and paradigms	'Space and time' to 'space-time-information' Formal or historical-philosophical
7	Design tools and sources	From letters and words to using computers From internal sources to external origins
8	Place and time	From place to placelessness From real to virtual but not actual spaces

Source Author

in design process classification. With the rise and rapid growth of new information and communication technologies, virtual spaces which are real but not actual have changed the way we experience the real and actual spaces. The comparison may open up a new horizon for designers.

Sometimes, it is not easy or even possible to draw a clear borderline to separate all design discourses. One may argue that all, or some, of the mentioned discourses are interwoven together. Designer roles or design scopes, participation, tools and non-place aspects of a design process may simply work together when the community participation occurs, for example, with web-based and virtual interactions in order to design with people and local authorities (Nourian and Rezaei 2006).

The dimensions of design process discussed here in this book through eight discourses might yet help readers to compare and find more relations between them all. Moreover, it may lead thinkers to discover connections between design theories and design techniques. Key texts examined here demonstrate an extensive literature in the field of design. It is possible to categorize design process theories in different types or scrutinize and add more texts into the tables for each design discourse. This may even change the way we define and understand the processes of design at all (Table 1).

Bibliography

Alexander C (1973) Notes on the synthesis of form. Harvard University Press, Harvard

Alexander C, Ishikawa S, Silverstein M, Jacobson M (1977) A pattern language. Oxford University Press, Oxford

Alexander C, Neis H, Anninou A, King I (1987) A new theory of urban design. Oxford University Press, New York

Analyzing the Concept of Urban Design (1999) TehranIran architecture and city-building research and studies center

Analyzing the Concept of Urban Design (1999a) TehranIran architecture and city-building research and studies center

Appleyard D (1976) Planning a pluralist city: Conflicting realities in Ciudad Guyana. MIT Press

Bacon E (1967) Design of cities. Thames & Hudson, London

Banerjee T (2001) The future of public space: beyond invented streets and reinvented places. J Am Plann Assoc 9–24

Bosselmann P (1998) Images in motion. In: Bosselmann P (ed) Representations of places: reality and realism in city design. University of California, Berkeley, pp 48–99

Carmona M (2001) Housing design quality: through policy, guidance, and review. Spon Press, London

Carmona M, Tiesdell S (2007) Urban design reader. Architectural Press, Oxford

Colquhoun A (1996) Typology and design method. In: Nesbitt K (ed) Theorizing a new agenda for architecture: an anthology of architectural theory 1965–1995. Princeton Architectural Press, New York, pp 248–257

Concepts in Architecture (1973), Mc Graw-Hill Book Company, New York

Cross N (1984) Developments in design methodology. Wiley, Hoboken

Davidof P (1965) Advocacy and pluralism in planning. J Am Plann 31

Design Cognition: Results from Protocol and Other Empirical from Protocol and Other Empirical Oxford Elsevier (2001)

Design in Architecture: Architecture and the Human Sciences (1973) Wiley, Chichester

Design Knowing and Learning: Cognition in Design Education (2001) Elsevier Science, UK

Design Methods (2007) Berlin Birkhauser architecture

Design Process (Decoding Analogy as A Major Method of Form and Space Producing) (2014b) Hoviat Shahr 71–80

Disordered Place: A critic on William H Whyte's Thoughts (2003b) City 2638–2642

Friedman J (1973) Retracking America: a theory of transactive planning. Doubleday-Anchor, NY

From Idea to Implementation: A participatory plan for the community in planning and urban design with applying ICT (2006) Fine Arts 35–44

M. Rezaei, *Reviewing Design Process Theories*,
SpringerBriefs in Architectural Design and Technology,
https://doi.org/10.1007/978-3-030-61916-9

GAUSA M (2003) Dispositions (composition > Position > Disposition. In GAUSA, GUALLART, MÜLLER, SORIANO, PORRAS, & MORALES. *Metapolis* Dictionary Adv Archit (p. 175). ACTAR Publishers, Barcelona

Gausa M, Guallart V, Muller W, Soriano F, Porras F, Morales J (2003) The metapolis dictionary of advanced architecture, city, technology and society in the information age. (S. Cros, Ed.), Actar, Barcelona

Gosling D, Maitland B (1984) Concepts of urban design. Martin's Press, St

Hatcher R, Fouad N, Grus C, Campbell L, Mccutcheon S, Leahy K (2013) Competency benchmarks: practical steps towards a culture of competence. Training Educ Prof Psychol 07

Hatchuel A (1996) Théories et modèles de la conception, Cours d'ingénierie de la conception. École des Mines de Paris, Paris

Hatchuel A, Weil B (2008) C-K design theory: an advanced formulation. Springer, Berlin

Hatchuel A, Weil B (2002) C-K theory: notions and applications of a unified design theory. In: Herbert simon international conference on "design sciences", Lyon, p 22

Hay L, Duffy AH, Mcteague C, Lyall L, Vuletic T, Grealy M (2017) A systematic review of protocol studies on conceptual design cognition: Design as search and exploration. Des Sci 3 (e10)

HCI reality—an 'Unreal Tournament'? (2007) Elsevier 737–743

How do you design? A compendium of models (2004) San Francisco Dubberly Design Office

Ideas that shaped Buildings (2003) Sabon by Graphic Composition, Inc., USA

Jones P, Gripaios P (2000) A review of the BURA awards for best practice in urban regeneration. Property Manage 218–229

Koberg D, Bagnall J (1990) The universal traveler. Crisp

Kumar V (2003) Innovation planning: modes, tools, uses. Chicago

Lawson B (1980) How designers think: The design process demystified. Architectural Press, Oxford

Maldonado T, Bonsiepe G (1964) Science and Design. ULM 10–29

Marzbani M, Jihad A, Rezaei M (2020) The sense of place: components and walkability. J Public Space 21–36

Moussavi F, Lopez D (2009) The function of form. ACTAR, Harvard Graduate of Design, London

Norberg-Schulz C (1979) Genius Loci: Towards a phenomenology of architecture. Rizzoli, Michigan

On the Analysis of Intuitive Design Processes (1970) Pittsburgh Pennsylvania

Parshall SA, William PM (2001) Problem seeking: An architectural programming primer. Wiley, Hoboken

Puglisi LP (1999) Hyper architecture: Spaces in the electronic age. Birkhauser, Berlin

Punter J (1992) The design dimension of planning: Theory, content and best practice for design policies

Punter J, Carmona M (1997) The design dimension of planning: theory, content and best practice for design policies. E & FN Spon, London

Rainer RM (2002) Letters on Cezann. Northpoint Press, New York

Rezaei M (2014) Design analytica: reviewing theories and concepts in contemporary design process of form and space. Islamic Aazad University, Central Tehran Branch, Tehran

Rezaei M (2003a) On the other paradigms: approaches to design: architects and urban designers. Archit Urbanism (70–71)

Rezaei M (2011) Reviews on Mozayeni's theories: passage. In: Mahmoodi MZ, Zaeimi F (eds) Immortal Wisdom. Tahan, Tehran, pp 154–157

Rezaei M (2004) The role of information and communication technologies in urban design with respect to the new urban development concepts. IAU- Science and Research Branch, Tehran, Iran

Rezaei M (2003d) Ultimate lesson: The path; review on manochehr mozayeni's theories. Tehran: Municipalities (Shahrdariha)

Rezaei M (2005) Urban design in the third millennium (the impacts of information and communication technology upon urban design). Faculty of Architecture and Urban Design, Sofeh, pp 4–27

Rezaei M (2013) Walkability criteria: the role of walkability in improving the sense of the place. Fine Arts J 18(4):15–24
Rittle HW, Webber MM (1973) Dilemmas in general theory of planning. Elsevier, pp 155–169
Salama A (1995) New trends in architectural education: designing the design studio. Tailored Text & Unlimited Potential Publishing, New Jersey
Schon (1983) The reflective practitioner. London
Sorkin M (2001) Some assembly required. University of Minnesota Press, London
Systematic Method for Designers (1965) Council of industrial design
The Application of Information and Communication Technology in Village Design Statement:(Proposing a new, participatory, and 'effective technique in improving districts) (2003c) Seminar on the application of ICT in Rural Areas 293–299 Tehran Science and Industry University
Townsend A (2003) Wired/Unwired: The urban geography of digital networks. MIT University, Cambridge, Massachusetts
Trancik R (1986) Finding lost space: theories of urban design. Wiley, Danvers
Ventury R, Scott Brown D, Izenour S (1972) Learning from las vegas. MIT Press, Cambridge
Vries M, Cross N, Grant DP (1993) Design methodology and relationships with science. Kluwer, Eindhoven

The manufacturer's authorised representative in the EU is Springer Nature Customer Service Centre GmbH, Europaplatz 3, 69115 Heidelberg, Germany. If you have any concerns regarding our products, please contact ProductSafety@springernature.com

Printed and bound by CPI Group (UK) Ltd, Croydon, CR0 4YY
30/09/2026
02225467-0006